NONLINEAR SEISMIC ANALYSIS AND DESIGN OF

REINFORCED CONCRETE BUILDINGS

This volume consists of papers presented at the Workshop on Nonlinear Seismic Analysis of Reinforced Concrete Buildings, Bled, Slovenia, Yugoslavia, 13-16 July 1992.

NONLINEAR SEISMIC ANALYSIS AND DESIGN OF REINFORCED CONCRETE BUILDINGS

Edited by

P. FAJFAR
University of Ljubljana, Slovenia

and

H. KRAWINKLER
Stanford University, USA

CRC Press
Taylor & Francis Group
Boca Raton London New York

CRC Press is an imprint of the
Taylor & Francis Group, an **informa** business

A TAYLOR & FRANCIS BOOK

CRC Press
Taylor & Francis Group
6000 Broken Sound Parkway NW, Suite 300
Boca Raton, FL 33487-2742

First issued in paperback 2019

ISBN-13: 978-1-85166-764-2 (hbk)
ISBN-13: 978-0-367-86394-4 (pbk)

British Library Cataloguing in Publication Data

Nonlinear Seismic Analysis and Design of
Reinforced Concrete Buildings
I. Fajfar, Peter II. Krawinkler, Helmut
693.54

Library of Congress CIP data applied for

PREFACE

This monograph is a compendium of invited papers that focus on the following two topics of much relevance to the seismic protection of reinforced concrete structures:
- (a) Energy concepts and damage models in seismic analysis and design
- (b) Analysis and seismic behavior of buildings with structural walls.

The papers are intended to set the stage for an assessment of the state-of-the-knowledge and the identification of future research and implementation needs on these two topics. They form the basis for discussions to take place at a workshop scheduled for summer 1992 and for recommendations to be developed and communicated to the research and design communities in a follow-up publication. It is hoped that both this monograph and the workshop will make a contribution to the achievement of the aims of the International Decade for Natural Disaster Reduction (IDNDR).

The two topics were selected because of their importance in seismic design and the need for an evaluation and dissemination of recent advances in these areas. It has long been recognized that energy input, absorption, and dissipation are the most fundamental quantities controlling seismic performance. Already in late fifties G.W. Housner proposed "a limit design type of analysis to ensure that there was sufficient energy-absorbing capacity to give an adequate factor of safety against collapse in the event of extremely strong ground motion". In 1960 John A. Blume, in his classical paper on the Reserve Energy Technique (2WCEE), states that "with the procedures outlined, the anomalies of a great deal of apparently baffling earthquake history can be explained as can the gap between elastic spectral data and the capacity to resist earthquakes". However, to this day, energy concepts have been ignored in earthquake resistant design because of apparent complexities in the quantification of energy demands and capacities and their implementation in the design process. The papers presented in the first part of this monograph illustrate how energy terms together with cumulative damage models can be utilized to provide quantitative information useful for damage assessment and design. It is hoped that a study of these papers leaves the reader with the impression that energy-based design is a viable concept, but it is also recognized that much more work needs to be done in order to simplify energy-based design to a level that makes it useful for design practice.

In many countries extensive use is made of structural walls (shear walls) to increase the strength and stiffness of lateral load resisting systems. Recent earthquakes have often indicated better performance of multistory buildings containing structural walls compared to buildings whose structural system consists of frames alone. Clearly, this observation cannot be generalized since seismic performance is affected greatly by wall layout, strength, and detailing, as well as by the primary deformation mode (bending versus shear). Although the great importance of walls in seismic performance has long been recognized, mathematical modeling of the nonlinear static and dynamic response of structures containing walls is only in the development stage. The second part of this monograph addresses important design and modeling problems for structural walls and buildings that rely on the participation o. walls in seismic resistance. The papers illustrate the complexity of the problems but also propose solution techniques intended to contribute to a more accurate prediction of the seismic behavior of buildings containing structural walls.

This monograph discusses selected issues of importance in the seismic design of reinforced concrete buildings. It lays no claim to providing final solutions to any of the problems investigated and probably raises more questions than it answers. Its purpose is to form a basis for discussion on the state-of-the-knowledge and research and implementation needs. Readers are encouraged to communicate their comments to the authors or the co-editors for consideration at the workshop for which these papers were written and which is scheduled for July 1992.

We are deeply indebted to the authors who have written original and thoughtful contributions to this monograph and have made commitments to participate in an international workshop in Bled near Ljubljana. This workshop was scheduled originally for June 1991 but has been postponed until summer of 1992. We are also much indebted to Elsevier Science Publishers Ltd who have generously agreed to make these publications available to the interested readers in a timely manner.

Sponsorship for the workshop for which these papers were written is provided by the U.S.-Yugoslav Joint Fund for Scientific and Technical Cooperation in conjunction with the U.S. National Science Foundation, the U.S. National Institute of Standards and Technology, the Ministry for Science and Technology of the Republic of Slovenia, and the Slovenian Academy for Sciences and Arts.

Peter Fajfar

Professor of Structural
and Earthquake Engineering
University of Ljubljana
Ljubljana, Slovenia

Helmut Krawinkler

Professor of Civil Engineering

Stanford University
Stanford, California, U.S.A.

CONTENTS

Behavior of buildings with structural walls

Energy concepts and damage models

ISSUES AND FUTURE DIRECTIONS IN THE USE OF AN ENERGY APPROACH FOR SEISMIC-RESISTANT DESIGN OF STRUCTURES

VITELMO V. BERTERO
Nishkian Professor of Structural Engineering
783 Davis Hall, University of California, Berkeley, CA 94720
Shimizu Corporation Visiting Professor at
Stanford University, Palo Alto, CA

CHIA-MING UANG
Assistant Professor, Civil Engineering Department
Northeastern University, Boston, MA 02115

ABSTRACT

This paper discusses the state-of-the-knowledge in the use of energy concepts in seismic-resistant design of structures emphasizing issues and future directions in the use of such concepts for proper establishment of design earthquakes. After a brief review of the nature of the earthquake problem, the need for improving the earthquake-resistant design of new structures and the proper upgrading of existing hazardous facilities is discussed. Emphasis is placed on the need and the difficulties of conducting nonlinear (inelastic) seismic design. The difference between design and analysis is pointed out, and the role of nonlinear analysis in the design process is discussed. The state-of-the-knowledge in the use of energy concepts in seismic-resistant design of new structures and particularly in the selection of proper (efficient) seismic upgrading of existing hazardous facilities is summarized. The importance of reliable estimation of the input energy of possible earthquake ground motions at the site of the structure in order to select the critical motion (i.e., to establish the proper design earthquake) is emphasized. The different engineering parameters that are needed for proper establishment of the design earthquake are discussed, concluding that while the input energy, E_I , is a reliable parameter for selecting the most demanding earthquake ground motion, it alone is not sufficient for proper design of the structure. For the sizing and detailing of a structure, it is necessary to specify the smoothed inelastic response spectra as well as the time history of the dissipated energy. Recommendations for research and development needs to improve the use of energy concepts in seismic-resistant construction are offered.

INTRODUCTION

STATEMENT OF THE ISSUES. It is well recognized that most human injury and economic loss due to moderate or severe earthquake ground motions are caused by the failure[1] of civil engineering facilities (particularly buildings), many of which were presumed to have been designed and constructed to provide protection against natural hazards. This has been dramatically confirmed during recent earthquakes around the world (the 1988 Armenia, 1989 Loma Prieta, 1990 Iran, and 1990 Philippines earthquakes). Therefore, one of the most effective ways to mitigate the destructive effects of earthquakes is to improve existing methods and/or develop new and better methods of designing, constructing and maintaining new structures and of repairing and upgrading (retrofitting) existing hazardous facilities, particularly buildings.

Although this paper will discuss only problems related to the seismic-resistant design of structures, it should be noted that, while a sound design is necessary, this is not sufficient to ensure a satisfactory earthquake-resistant structure. The seismic response of the structure depends on the state of the whole soil-foundation-superstructure and nonstructural components system when earthquake shaking occurs, i.e., response depends not only on how the structure has been designed and constructed, but on how it has been maintained up to the time that the earthquake strikes. A design can only be effective if the model used to engineer the design can be and is constructed and maintained. Although the importance of construction and maintenance in the seismic performance of structures has been recognized, insufficient effort has been made to improve these practices (e.g., supervision and inspection) [1].

In an attempt to realize the above mentioned improvements, the authors and their research associates have carried out a series of studies examining the problems encountered in improving earthquake-resistant design of new structures and the development of more reliable approaches to the seismic upgrading of existing hazardous facilities. Because the fundamental earthquake ground motion data required to conduct reliable vulnerability assessment of existing structures and facilities and then to develop efficient strategies for seismic upgrading of hazardous structures is the same as that required for earthquake-resistant design of new structures, only this last case will be discussed herein with special emphasis on building structures.

The state-of-the-art and the state-of-the-practice of earthquake-resistant design and construction of buildings have been reviewed in a series of recent publications by the author and his colleagues [Refs. 2-4]. The importance of a number of problems that have been under study and mentioned in these reviews has recently been confirmed by: the ground motions recorded during two major earthquakes in 1985 (March 3rd in Chile and September 19th in Mexico) and the 1989 Loma Prieta earthquake; the results obtained from the processing of these records; the performance of the structures, particularly buildings during the above and other recent earthquakes; and the results of integrated analytical and experimental studies that have recently been conducted. To recognize the importance as well as the difficulties involved in the solution of the general issues (problems) encountered in the seismic-resistant design of structures, it is convenient to briefly review these problems.

Overview of Special Problems Encountered in the Design of Earthquake-Resistant Structures. To conduct efficient earthquake-resistant design of a facility (for example, a building), it is necessary to predict reliably the mechanical (dynamic) behavior of the

---

[1]The term failure is used herein not only to represent physical collapse, but also any serious structural a nonstructural damage which can jeopardize human life and/or the function of the facility.

whole soil-foundation-superstructure and nonstructural components of the building system. The general problems involved in predicting the seismic response of a building are symbolically defined and schematically illustrated in Fig. 1.

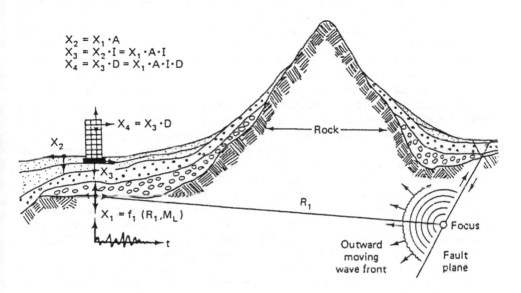

$$X_2 = X_1 \cdot A$$
$$X_3 = X_2 \cdot I = X_1 \cdot A \cdot I$$
$$X_4 = X_3 \cdot D = X_1 \cdot A \cdot I \cdot D$$

$X_4 = X_3 \cdot D$

X_2

Rock

X_3

R_1

$X_1 = f_1 (R_1, M_L)$

Focus

Outward moving wave front

Fault plane

EARTHQUAKE OF MAGNITUDE M_L

Fig. 1 Illustration of Problems and Factors Involved in Predicting the Seismic Response of a Building

The first problem is to estimate accurately the ground motion at the foundation of the building, X_3. For an earthquake of specified magnitude, M_L, and focal distance, R_1, it is analytically feasible to estimate the base rock motion at the given site of the building, X_1, if the fault type is known $[X_1 = f(R_1, M_L)]$. Prediction of X_3, however, must account for the effects of the soil layers underlying and/or surrounding a building. These effects can be classified in two groups: One is related to the influence of the dynamic characteristics of the different soil layers during the transmission of X_1 to the free ground surface, indicated in Fig.1 by an attenuation or amplification factor, A $[X_2 = A \cdot X_1]$; the other is related to the interaction between structure and soil foundation, symbolically represented by a factor I. There are presently large uncertainties regarding the realistic values of A and I, and major errors could be introduced by trying to quantify these two factors using available analytical techniques. Even if X_1 could be predicted with engineering accuracy, attempts to quantify the influence of soil conditions on X_1 to attain X_2 and X_3 would result in a wide range of predicted values. Soil behavior can be very sensitive to the intensity of the seismic waves, as well as to the rate of straining they could induce. Thus, the analyst or designer would not rely exclusively on results obtained from a single deterministic analysis. Bounds on the possible variations in A and I should also be considered.

The second problem is to predict the deformation, X_4, from shaking at the foundation, X_3 by a dynamic operator, D. Although this is a simple expression, the uncertainties involved in realistically estimating X_3 and D give rise to serious difficulties in obtaining an accurate numerical evaluation of X_4.

Even if it were possible to predict the mechanical characteristics of a building, there remain many uncertainties in establishing the six components of the critical X_3, and attempts to predict the response (strength and deformation) of the building should consider the complete range, or at least the bounds of the dynamic characteristics of the possible excitations, X_3. Owing to these uncertainties, the ideal solution would be a conservative one which would identify the critical excitation, X_3, for the given building. Although it is easy to define this critical X_3 as that which drives a structures to its maximum response, its specific quantification is more complicated. Quantification is feasible for elastic response, but is complicated for cases involving nonlinear response.

The precise evaluation of X_4 at any point in a structure requires that its three translational and three rotational components be established. Furthermore, the prediction of X_4 for a particular building to a specific ground motion depends on the combined effect and dynamic characteristics of all excitations acting on the building. Usually the main excitations on a structures during a severe earthquake are due to: (1) Gravity forces, G(t), with the associated effects of volumetric changes produced by creep of the structural material, especially in concrete; (2) changes in environmental conditions, E(t), such as stresses produced by variations in temperatures; and (3) at least the three translational components of the foundation shaking, $X_3(t)$.

As shown in eq. (1a), the dynamic characteristics of the whole system, which can change continuously as the structures is deformed into its inelastic range, can be summarized by denoting them as the instantaneous values of: (1) Mass, M(t); (2) damping coefficient, $\xi(t)$; and (3) resistance function, (R versus X_4)(t). They can also be represented as illustrated in eq.(1b) by the instantaneous values of: (1) Fundamental period, T(t); (2) damping coefficient, $\xi(t)$; (3) yielding strength, $R_y(t)$; and (4) energy absorption and dissipation capacity as denoted by instantaneous available ductility, $\mu(t)$, which is a function of $X_4(t)$.

$$X_4(t) = F\{[G(t), \Delta E(t), X_3(t)], \qquad [M(t), \xi(t), (R \text{ vs. } X_4)(t)]\} \qquad (1a)$$

$$X_4(t) = F\{\underbrace{[G(t), \Delta E(t), X_3(t)]}, \qquad \underbrace{[T(t), \xi(t), R_y(t), \mu(t)]}\} \qquad (1b)$$

**dynamic characteristics   dynamic characteristics of
of excitations   whole soil-building system**

Analysis of the parameters in eqs. (1a) and (1b) indicates the magnitude of the problems involved in predicting response to earthquake ground motions. The first problem is that to predict X_4, X_4 must be known. Another problem is that all such parameters are functions of time, although the gravity forces and environmental conditions tend to remain nearly constant for the duration of an earthquake. It should be noted that the value of $\Delta E(t)$ represents more than just the stresses induced by environmental changes that occur during the critical earthquake ground motion, X_3; it also accounts for stress and strain existing at the time of the earthquake due to (1) previous thermal changes or shrinkage, which cause residual stress or distress, and deterioration from aging and corrosion; (2) degradation in strength and stiffness caused by previous exposure to high winds, fires, or earthquakes; (3) strength and stiffness caused by alternation, repair or strengthening. Since any one of these conditions can significantly affect structural response, factors that must be considered in determining the strength and deformation capacities include the variations in loading and environmental histories during the service life of the building and their effects on the condition of the structure at the time of the occurrence of the extreme environment. To this end, it should be noted that $E(t)$ also affects $(R$ versus $X_4)(t)$.

Another difficulty is that in the case of predicting the response, $X4(t)$, to the extreme (safety) ground motions, this usually involves nonlinear (inelastic) response. Therefore, it is not possible to apply the principle of superposition and solve independently the problem for each of the different excitations and then superimpose their solution. This is one of the main reasons why in practice designers prefer to reduce (simplify) the prediction of the seismic response of a structure and, therefore, limit its design to the linear elastic range of the actual response.

Differences Between Analysis and Design, and Between Design and Construction. A preliminary structural design should be available to conduct linear elastic and nonlinear (inelastic) analyses of the soil-foundation-superstructure model(s). To recognize clearly the differences between analysis and design, and at the same time identify problems inherent in the design of earthquake-resistant structures, it is convenient to analyze the main steps involved in satisfying what can be called the **basic design equation:**

DEMAND	$\leq$	SUPPLY
on		of
Stiffness		Stiffness
Strength		Strength
Stability		Stability
Energy absorption & energy dissipation capacities		Energy absorption & energy dissipation capacities

Evaluation of the **demand** and prediction of the **supply** are not straightforward, particularly for earthquake-resistant buildings. Determination of the **demand**, which usually is done by numerical analyses of mathematical models of the entire soil-foundation-building system, depends on the interaction of this system as a whole and the different excitations that originate from changes in the system environment and of the **intrinsic interrelation between the demand and supply itself.**

In the last three decades our ability to analyze mathematical models of buildings when subjected to earthquake ground shaking has improved dramatically. Sophisticated computer programs have been developed and used in the numerical analysis of the linear as well as nonlinear seismic response of three-dimensional mathematical models of the bare structure of a building to certain assumed earthquake ground motions (**earthquake input**). The opportunity is ripe to take advantage of these improvements in analysis in the seismic design of structures. In general, however, these analyses have failed to predict the behavior of real buildings, particularly at ultimate limit states. As a consequence of this and due to the lack of reliable models to predict **supplies to real structures**, there has not been a corresponding improvement in the design of earthquake-resistant structures. There is an urgent need to improve mathematical modeling of real facilities. This requires integrated analytical and experimental research.

The proportioning (sizing) and detailing of the structure elements of a building are usually done through equations derived from the theory of mechanics of continuous solids or using empirical formulae. Except in the case of pure flexure, a general theory with reliable equations that can accurately predict energy absorption and dissipation capacities of structural elements and of the so-called nonstructural elements, and therefore of real buildings, has not been developed. Improving this situation will require integrated analytical and experimental research in the field (through intensive instrumentation of buildings) and in the experimental laboratory through the use of pseudo-dynamic and/or earthquake simulator facilities.

The information needed to improve prediction of earthquake responses of structures, and therefore necessary to improve their design, can be grouped under the following three basic elements: **Earthquake input, demands on the structure, and supplied capacities of the structure.** The authors believe that a promising approach for improving the solution of the problems involved with these three elements is through the use of energy concepts [6]. To review the state-of-the-art in the use of these concepts is one of the main objectives of this paper. Because of space limitations, this paper will attempt to focus on the state-of-the-art in the use of such an energy approach only to solve the problem of proper selection of the earthquake input.

<u>Earthquake Input: Specification of Design Earthquakes and Design Criteria</u>. The design earthquake depends on the design criteria, i.e., **the limit state controlling the design.** Conceptually, the design earthquake should be that ground motion that will drive a structure to its critical response. In practice, the application of this simple concept meets with serious difficulties because, first, there are great uncertainties in predicting the main dynamic characteristics of ground motions that have yet to occur at the building site, and , secondly, even the critical response of a specific structural system will vary according to the various limit states that could control the design.

Until a few years ago seismic codes have specified design earthquakes in terms of a building code zone, a site intensity factor, or a peak site acceleration. Reliance on these indices, however, is generally inadequate, and methods using **ground motion spectra (GMS)** and **Smoothed Elastic Design Response Spectra (SEDRS)** based on **Effective Peak Acceleration (EPA)** have been recommended [2-5]. While this has been a major improvement conceptually, great uncertainties regarding appropriate values for EPA and GMS, as well as other parameters that have been recommended to improve this situation, persist [5 & 7]. The authors believe that a promising engineering parameter for improving selection of proper design earthquakes is through the concept of **Energy Input, E_I, and associated parameters.** The use of this concept and associated parameters is the main subject of this paper which has the following objectives.

OBJECTIVES. The main objectives of this paper are: First, to discuss the state-of-the-knowledge in the use of energy concepts in seismic-resistant design of structures with emphasis on the proper establishment of the design earthquakes through the use of E_I and associated parameters; and, secondly, to point out the main issues and future directions in the use of such an energy approach.

STATE-OF-THE-KNOWLEDGE IN THE USE OF ENERGY CONCEPTS

GENERAL REMARKS. Traditionally, displacement ductility has been used as a criterion to establish **Inelastic Design Response Spectra (IDRS)** for earthquake-resistant design of buildings [8]. The minimum required **strength** (or capacity for lateral force) of a building is then based on the selected IDRS. As an alternative to this traditional design approach, an energy-based method was proposed by Housner [9]. Although estimates have been made of input energy to Single Degree of Freedom Systems, SDOFS, [10] and even to **Multi-Degree of Freedom Systems, MDOFS,** (steel structures designed in the 60's for some of the existing recorded ground motions) [11], it is only recently that this approach has gained extensive attention [12]. This design method is based on the premise that the **energy demand** during an earthquake (or an ensemble of earthquakes) can be predicted and that the **energy supply** of a structural element (or a structural system) can be established. A satisfactory design implies that the energy supply should be larger than the energy demand.

To develop reliable design methods based on an energy approach, it is necessary to derive the energy equations. Although real structures are usually **MDOFS**, to facilitate the analysis and understanding of the physical meaning of the energy approach, it is convenient to first derive the energy equations for **SDOFS** and then to derive these equations for **MDOFS**.

DERIVATION OF ENERGY EQUATIONS: Linear Elastic-Perfectly Plastic SDOFS.
In Ref. 13 is a detailed discussion of the derivation of the following two basic energy equations starting directly from the eq. (2) for a given viscous damped SDOFS subjected to an earthquake ground motion

$$m\ddot{v}_t + c\dot{v} + f_s = 0 \tag{2}$$

where: m = mass; c = viscous damping constant; f_s = restoring force (if k = stiffness, f_s = kv for a linear elastic system); $v_t = v + v_g$ = absolute (or total) displacement of the mass; v = relative displacement of the mass with respect o the ground; and v_g = earthquake ground displacement.

Derivation of "Absolute" Energy Equation. Integrating Eq. 2 with respect to v from the time that the ground motion excitation starts and considering that $v = v_t - v_g$ it can be shown that

$$\frac{m(\dot{v}_t)^2}{2} + \int c\dot{v}\,dv + \int f_s\,dv = \int m\ddot{v}_t\,dv_g \tag{3}$$

$$E_K + E_\xi + E_a = E_I \tag{4}$$

| "Absolute" Kinetic Energy | Damping Energy | Absorbed Energy | = | "Absolute" Input Energy |

Considering that E_a is composed of the recoverable Elastic Strain Energy, E_s, and of the irrecoverable Hysteretic Energy, E_H, eq. (4) can be rewritten as

$$E_I = E_K + E_S + E_\xi + E_H \tag{5}$$

The E_I is defined as the **"Absolute Input Energy"** because it depends on the absolute acceleration, $\ddot{v}_t$. Physically, it represents the inertia force applied to the structure. This force, which is equal to the restoring force plus damping force [see eq. (2) and Fig. 2], is the same as the total force applied to the structure foundation. Therefore, E_I represents the work done by the total base shear at the foundation on the foundation displacement, v_g.

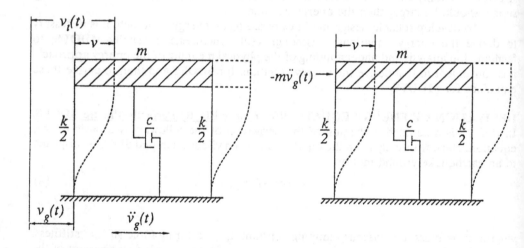

(a) Moving Base System　　　　　　　(b) Equivalent Fixed-Base System

Fig. 2 Mathematical Model of a SDOFS subjected to an Earthquake Ground Motion

Derivation of "Relative" Energy Equation.　Considering that Eq. (2) can be rewritten as

$$m\ddot{v} + c\dot{v} + f_s = -m\ddot{v}_g \tag{6}$$

and the structural system of Fig. 2(a) can be conveniently treated as the equivalent system in Fig. 2(b) with a fixed base and subjected to an effective horizontal dynamic force of magnitude, $m\ddot{v}_g$. Integrating Eq. (6) with respect to v, the following eq. can be obtained:

$$\frac{m(\dot{v})^2}{2} \quad + \quad \int c\dot{v}dv \quad + \quad \int f_s dv \quad = \quad -\int m\ddot{v}_g dv \qquad (7)$$

$$E_K' \quad + \quad E_\xi \quad + \quad E_a \quad = \quad E_I' \qquad (8)$$

"Relative" Kinetic Energy	Damping Energy	Absorbed Energy	"Relative" Input Energy

As $E_a = E_s + E_H$, Eq. (8) can be rewritten as

$$E_I' = E_K' + E_s + E_\xi + E_H \qquad (9)$$

The E_I' that is defined as the "Relative" Input Energy represents the work done by the static equivalent internal force ($-m\ddot{v}_g$) on the equivalent fixed-base system; that is, it neglects the effect of the rigid body translation of the structure.

Difference Between Input Energies from Different Definitions. Ref. 13 discusses in detail the differences between the values of the input energies E_I and E_I'. Although the profiles of the energy time histories calculated by the absolute energy equation (3) differ significantly from those calculated by the conventional relative equation, eq. 7, the maximum values of E_I and E_I' for a constant displacement ductility ratio are very close in the period range of practical interest for buildings which is 0.3 to 5.0 secs.

Comparison of E_I with the Maximum Input Energy that is Stored in a Linear Elastic SDOFS. For a linear elastic SDOFS the maximum input energy that is stored is given by

$$E_{IS} = \frac{1}{2} m (S_{pv})^2 \qquad (10)$$

where S_{pv} is the linear elastic spectral pseudo-velocity.

This E_{IS} has been used by some researchers as the energy demand for an inelastic system. In Ref. 13 it is shown that E_{IS} may significantly underestimate the E_I for an inelastic system. In this reference, it is also shown that, except for highly harmonic ground motions (like the recorded one at SCT in Mexico City during the 1985 earthquake), the E_I for a constant ductility ratio can be predicted reliably by the elastic input energy spectra using Iwan's procedure [14] which takes into consideration the effect of increasing damping ratio and natural period.

Input Energy to MDOFS. The E_I for an N-story building can be calculated as follows [13]:

$$E_I = \int (\sum_{i=1}^{N} m_i \ddot{v}_{ti}) dv_g \qquad (11)$$

Where: m_i is the lumped mass associated with the i-th floor, and $\ddot{v}_{ti}$ is the total acceleration at the i-th floor. In other words, E_I is the summation of the work done by the total inertia force $(m_i v_{ti})$ at each floor through the ground displacement v_g. Analysis of results obtained from experiments conducted on medium rise steel dual systems indicates that the E_I to a multi-story building can be estimated with sufficient practical accuracy by calculating the E_I of a SDOFS using the fundamental period of the multi-story structure.

ADVANTAGES OF USING ENERGY CONCEPTS IN SEISMIC DESIGN OF STRUCTURES.   Equation (5) can be rewritten as

$$E_I = \quad E_E \quad\quad + \quad\quad E_D \tag{12a}$$

$$E_I = \overbrace{E_K + E_S} \quad + \quad \overbrace{E_\xi + E_H} \tag{12b}$$

where E_E can be considered as the stored elastic energy and E_D the dissipated energy. Comparing this equation with the design equation, it becomes clear that E_I represents the **demands**, and the summation of $E_E + E_D$ represents the **supplies**. This eq. (12a) points out clearly to the designer that to obtain an efficient seismic design, the first step is to have a good estimate of the E_I for the critical ground motion. Then the designer has to analyze if it is possible to balance this demand with just the elastic behavior of the structure to be designed or will it be convenient to attempt to **dissipate** as much as possible some of the E_I , i.e., using E_D. As revealed by eq. (12b), there are three ways of increasing E_D: One is to increase the linear viscous damping, E_ξ;   another, is to increase the hysteretic energy, E_H; and the third is a combination of increasing E_ξ and E_H. At present it is common practice to just try to increase the E_H as much as possible through inelastic (plastic) behavior which implies damage of the structural members. Only recently it has been recognized that it is possible to increase significantly the E_H and control damage through the use of **Energy Dissipation Devices.**

If technically and/or economically it is not possible to balance the required E_I through either E_E alone or $E_E + E_D$, the designer has the option of attempting to control (decrease) the E_I to the structure.   This can be done by **Base Isolation Techniques.**   A combination of controlling (decreasing) the E_I by base isolation techniques and increasing the E_D by the use of energy dissipation devices is a very promising strategy not only for achieving efficient seismic-resistant design and construction of new structures, but also for the seismic upgrading of existing hazardous structures [15]. To reliably use this energy approach, it is essential to be able to select the critical ground motion (design earthquake), i.e., that which controls the design; in other words, the ground motion that has the largest damage potential for the structure being designed. Although many parameters have been and are being used to establish design earthquakes, most of them are not reliable for assessing the damage potential of earthquake ground motions. As mentioned in the Introduction, a promising parameter for assessing damage potential of these motions is the E_I [6]. However, as it will be discussed below, this parameter alone is not sufficient to evaluate (visualize) the E_D (particularly E_H) that has to be supplied to balance the E_I for any specified acceptable damage. Additional information is needed.

INFORMATION NEEDED TO CONDUCT RELIABLE
SEISMIC-RESISTANT DESIGN

GENERAL REMARKS. It has been pointed out previously that the first and fundamental step in seismic-resistant design of structures is the reliable establishment of the design earthquakes. This requires a reliable assessment of the damage potential of all the possible earthquake ground motions that can occur at the site of the structure. An evaluation of the different parameters that have been and are still used is offered in Ref. 6. Currently, for structures that can tolerate a certain degree of damage, the **Safety or Survival-Level Design Earthquake** is defined through **Smoothed Inelastic Design Response Spectra, SIDRS.** Most of the SIDRS that are used in practice (seismic codes) have been obtained directly from **SEDRS**, through the use of the **displacement ductility** ratio, μ, **or reduction factors, R.** The validity of such procedures has been questioned, and it is believed that at present such SIDRS can be obtained directly as the mean or the mean plus one standard deviation of the **Inelastic Response Spectra, IRS,** corresponding to all the different time histories of the severe ground motions that can be induced at the given site from earthquakes that can occur at all of the possible sources affecting the site [7].

While the above information **is necessary** to conduct reliable design for safety, i.e., to avoid collapse and/or serious damage that can jeopardize human life, **it is not sufficient.** Although the IRS takes into account the effects of duration of strong motion in the required strength, these spectra do not give an appropriate idea of the amount of energy that the whole facility system will dissipate through hysteretic behavior during the critical earthquake ground motion. They give only the value of maximum global ductility demand. In other words, **the maximum global ductility demand by itself does not give an appropriate definition of the damage potential of ground motions.** In Ref. 6 the authors have shown that a more reliable parameter than those presently used in assessing damage potential is the E_I. As is clearly shown by eqs. 3 and 4, this damage potential parameter depends on the dynamic characteristics of both the shaking of the foundation and the whole building system (soil-foundation-superstructure and nonstructural components). Now the question is: Does the use of the SIDRS for a specified global μ and the corresponding E_I of the critical ground motion give sufficient information to conduct a reliable seismic design for safety?

Although the use of E_I can identify the damage potential of a given ground motion and, therefore, permits selection, amongst all the possible motions at a given site, of that which will be the critical one for the response of the structure, it does not provide sufficient information to design for safety level. From recent studies [7 & 13] it has been shown that the energy dissipation capacity of a structural member, and therefore of a structure, depends upon both the loading and deformation paths. Although the energy dissipation capacity under monotonic increasing deformation may be considered as a lower limit of energy dissipation capacity under cyclic inelastic deformation, the use of this lower limit could be too conservative for earthquake-resistant design. This is particularly true when the ductility deformation ratio, say μ, is limited, because of the need to control damage of nonstructural components or other reasons, to low values compared to the ductility deformation ratio reached under monotonic loading. Thus, effort should be devoted to determining experimentally the energy dissipation capacity of main structural elements and their basic subassemblages as a function of the maximum deformation ductility that can be tolerated, and the relationship between energy dissipation capacity and loading and/or deformation history.

From the above studies, it has also been concluded that damage criteria based on the simultaneous consideration of E_I and μ (given by SIDRS), and the E_H (including **Accumulative Ductility Ratio, μ_a , and Number of Yielding Reversals, NYR**) are promising for defining rational earthquake-resistant design procedures. The need for considering all of these engineering parameters rather than just one will be justified below by a specific example. From the above discussion, it is clear that when significant damage can be tolerated, the search for a single parameter to characterize the ground motion or the design earthquake for safety is doomed to fail.

IMPORTANCE OF SIMULTANEOUSLY CONSIDERING THE E_I , IDRS, AND E_H (INCLUDING μ_a AND NYR) FOR DEFINING THE SAFETY-LEVEL DESIGN EARTHQUAKE. Figures 3-7 permit comparison of the values of these different engineering parameters for two recorded ground motions, San Salvador (SS) and Chile (CH); Table 1 summarizes approximate maximum values for these parameters corresponding to each of these two different recorded ground motions. The importance and, actually, the need for simultaneously considering all the above parameters in selecting the critical ground motions and, therefore, for defining the safety-level design earthquake, is well illustrated by analyzing the values of these parameters for these two records.

<u>San Salvador (SS) vs. Chile (CH) Records.</u> From analyses of the values of Peak **Ground Acceleration (PGA), Effective Peak Acceleration (EPA),** and **Effective Peak Velocity (EPV)** (given in Table 1) which are values presently used to define the seismic hazard zoning maps, it might be concluded that the damage potential of these ground motions is quite similar. One can arrive at a similar conclusion if the values of the required **Yielding Strength Coefficient, C_y = V_y /W** , for different values of μ are compared or, in other words, if the IRS for different μ are compared (Fig. 3). However, a completely different picture is obtained when the values of the E_I , E_H, μ_a, and NYR for different values of μ are compared. The E_I for the CH record can be as much as 5 times the E_I for the SS record (Fig. 4). The E_H [represented by the equivalent hysteretic velocity, V_H = $(2E_H /m)^{1/2}$, in Fig. 5] for the CH record is more than 3 times the E_H for the SS record when the period, T, is about 0.5 secs. and nearly 2 times when the T varies from 0.5 secs. up to 1.5 secs. The μ_a for the CH record are 2 to 4 times higher than those of the SS record (Fig. 6). The NYR for the CH record and for a μ = 6 and T < 0.5 seconds are more than 10 times the NYR for the SS record (Fig. 7). For a μ = 4 and T $\geq$ 0.5, the NYR for the CH record are more than 5 times those of the SS record.

From the above comparison, it is clear that the damage potential of the CH recorded ground motion is significantly (at least 3 times) greater than that of the SS record in spite of the fact that PGA, EPA, EPV, ERS (IRS for μ = 1) and even the IRS for different values of μ are very similar. Thus, the importance of evaluating the E_I and E_H (represented herein by V_H, μ_a and NYR spectra) which are functions of the duration of strong ground motions, t_d , becomes very clear. While the t_d for the CH records is 36 secs., the t_d for the SS record is only 4.3 secs.(see Table 1). The importance of t_d in judging damage control is discussed in Ref. 7. While the above spectra are very helpful in preliminary design, for the final design (detailing of members), the ideal would be to have the time history of the E_H, i.e., the time history of the load-deformation relationship of the designed structure.

Table 1. Parameters Corresponding to the Chile (CH) and San Salvador (SS) Earthquake Ground Motions

EQ GROUND MOTION PARAMETERS / EQ RECORD	PGA (g)	EPA (g)	EPV (in/s)	t_d (secs)	$\mu = 2$				$\mu = 4$				$\mu = 6$			
					C_y	E_I/m $\frac{in^2}{sec^2}$	μ_a	NYR	C_y	E_I/m $\frac{in^2}{sec^2}$	μ_a	NYR	C_y	E_I/m $\frac{in^2}{sec^2}$	μ_a	NYR
CHILE (CH)	0.67	0.57	16	35.8	0.95	11,200	11	28	0.70	9,600	33	53	0.67	8,800	133	122
SAN SALVADOR (SS)	0.69	0.54	17	4.3	1.04	2,400	5	6	0.69	1,900	12	9	0.69	1,700	28	9

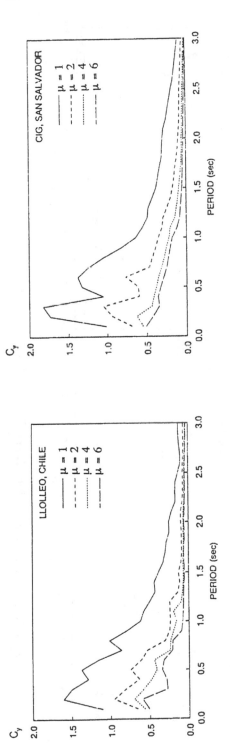

Fig. 3 Yielding Strength Spectra (C_y) for CH and SS Records (5% Damping)

16

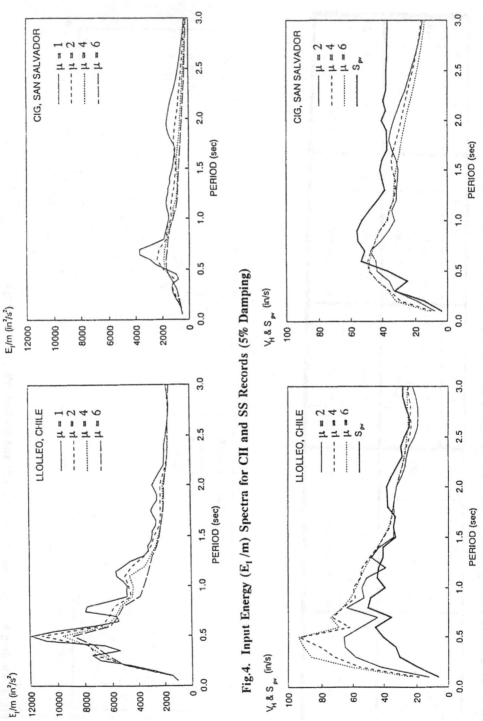

Fig.4. Input Energy (E_I/m) Spectra for CH and SS Records (5% Damping)

Fig.5. Hysteretic Energy Equivalent Velocity (V_H) Spectra for CH and SS Records (5% Damping)

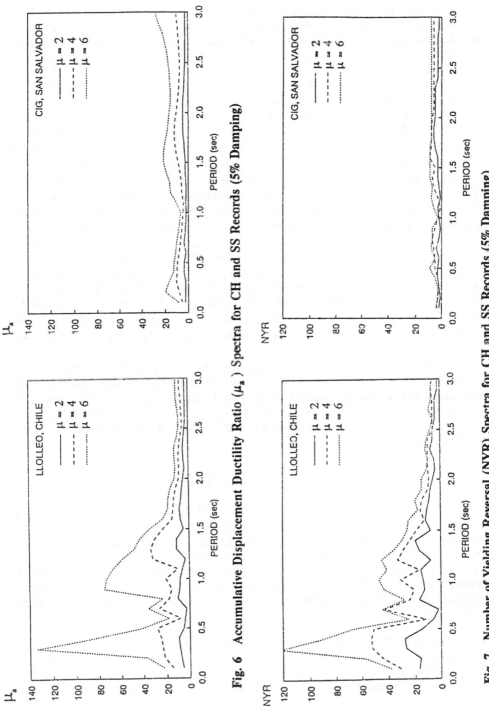

Fig. 6 Accumulative Displacement Ductility Ratio (μ_a) Spectra for CH and SS Records (5% Damping)

Fig. 7 Number of Yielding Reversal (NYR) Spectra for CH and SS Records (5% Damping)

The assembly of all the above spectra and time histories can be considered the **ideal information** for making reliable decisions regarding the critical earthquake ground motions and, therefore, for reliable establishment of design earthquake and design criteria. Thus, the gathering of this basic information should be pursued for research in order to improve seismic codes as well as for the design of important facilities. It should be noted that all of the above spectra can be computed by the engineer if he/she is provided with the time history of all possible ground motions than can occur at the site of the structure.

It has to be recognized that, for practical preliminary design of most standard facilities, it will be convenient to specify the minimum possible information to keep it simple. It is believed that, for a given structural site, this minimum could be the E_I and the SIDRS of all the possible ground motions as that site. The E_I would permit selection of the type of critical ground motion, i.e., the one that will induce the largest damage. The SIDRS, corresponding to the type of critical ground motion, can be used to conduct the preliminary design of the structure. Once a preliminary design is completed, it will be possible to obtain all the other information, i.e., the E_H, μ_a and NYR for different μ, from nonlinear, dynamic time history analyses, taking advantage of the significant advances achieved in the development of computer programs for such analysis. This will permit checking the adequacy of the preliminary design. While a nonlinear analysis of the preliminary design using a static approach (i.e., equivalent static lateral force) can give an idea of the strength and deformation capacities as well as a lower limit of the available E_H and therefore it should be used if no time history of the critical ground motions is possible, this type of analysis will not supply any information regarding the μ_a or NYR or the sequence of damage.

From the above discussion, it becomes clear that, if future codes perpetuate simple procedures for seismic design specifying only smoothed strength response spectra, it will be necessary to place more stringent limitations on the type of structural systems that could be used and on how such procedures can be applied, and to have very conservative regulations in the sizing and detailing for ductility and in the maximum acceptable deformations.

CONCLUSIONS AND RECOMMENDATIONS

CONCLUSIONS. Review of the state-of-the-knowledge in seismic-resistant design of structures reveals that among the several issues that exist the following two are very important: First, what are the earthquake effects, and in particular for any selected site of a structure, what are the ground motions against which the structure has to be designed? And secondly, how should the design be conducted to resist such earthquake effects? From the above discussions and the analyses of the results obtained in the studies published in the references cited in this paper regarding these two main issues, the following observations can be made:

(1) The application of energy concepts through the use of energy equations has the advantage that it guides (indicates) the designer through the different alternatives at his disposal to find an efficient (technical and economical) design. It encourages and guides the designer in the proper application of recent developments in the use of techniques

in base isolation and energy dissipation devices.

(2)   Of the two energy equations derived from analysis of response of SDOFS, the use of the **"absolute"** energy equation rather than the **"relative"** energy equation has the advantage that the physical energy input is reflected.

(3)   The absolute and relative input energies for a constant displacement ductility ratio are very close in the period range of practical interest, namely 0.3 to 5 secs.

(4)   For certain types of earthquake ground motion, the absolute input energy spectra are sensitive to the variation of the ductility ration, μ.

(5)   The use of the stored energy in a linear elastic SDOFS, $E_{IS} = m(S_{pv})^2/2$, as a measure of the energy demand for an inelastic system can significantly underestimate the E_I.

(6)   The E_I for a constant ductility ratio can be predicted reliably based on the use of Elastic Input Energy Spectra provided that the increased damping ratio and natural period proposed by Iwan are used. An exception to this is when the ground motion is of the type (highly harmonic) recorded at the SCT station in Mexico City during the 1985 Mexico earthquake.

(7)   For certain types of multi-story buildings (i.e., MDOFS), their E_I can be estimated with sufficient reliability from the E_I of the SDOF using the fundamental period of the MDOFS.

(8)   For proper establishment of design earthquakes and design criteria, it is necessary to have a reliable assessment of the damage potential of each of the different ground motions that can occur at the site of the structure. The different parameters used currently in practice (codes) are inadequate in assessing the damage potential of an earthquake ground motion. The E_I is a reliable parameter in selecting the most demanding earthquake; however, it **alone** is not sufficient for conducting reliable design, sizing and detailing when damage can be tolerated.

(9)   The conventional inelastic response spectra, based on a constant μ, cannot be used **alone** as a parameter for judging the damage potential of the earthquake ground motions and thus for establishing the design earthquakes. These spectra do not reflect the possibility of high energy dissipation demand for earthquakes with long duration of strong motion.

(10)   For a given structure site, the best parameters for selecting critical earthquake ground motions at the safety or survivability level are the E_I and E_H spectra corresponding to all the possible earthquake ground motions that can occur (or have been recorded) at the selected site. While E_I represents the total energy demand, the E_H is directly related to the damage (inelastic deformations) that can be expected.

(11)   The E_H spectra are generally in close agreement with the spectra for the energy stored in a linear elastic SDOFS, $m(S_{pv})^2/2$; however, this elastic stored energy may

underestimate significantly the E_H for structures subjected to long duration of strong motions as those recorded during the 1985 Chile and Mexico earthquakes.

(12) For proper selection of the structural system to be used and particularly for the sizing and detailing of the structural members, the use of the E_I, IRS and E_H spectra is not sufficient. The energy dissipation capacity of a structure is dependent on the deformation path (i.e., the deformation time history). Thus, it is necessary to have information about: The accumulative deformation ductility ratio, μ_a, and the number of yielding reversals, NYR. The μ_a and NYR spectra are not enough, however, by themselves; what is needed is the deformation time history of the critical regions of the selected structural system when subjected to the critical ground motions.

(13) It looks as if any attempt to base seismic design of structures on a design earthquake developed by only one engineering parameter is doomed to fail. But if the present building code philosophy of maintaining as simple a code as possible persists and it is also desired to base seismic design on the formulation of just a design response spectra, it is suggested that this be accomplished by specifying SIDRS for different levels of μ and for different types of site conditions rather than to specify SEDRS and R factors and site coefficients, S. This should be done together with very strict limitations on the type of structural systems to be used and very stringent requirements on the sizing and detailing of the structural components to achieve the largest possible ductility deformation ratio that can be economically attained so that the structure can economically supply a large E_H. Even if these restrictions are specified, it would be desirable that the codes require nonlinear (inelastic) analyses of the preliminarily design structure at least under static lateral forces. Ideally, it should be done using time history analyses.

RECOMMENDATIONS: To improve solutions of the existing issues in the use of energy concepts and their application in practice, the following needs are identified.

　　Research Needs.　　Most of the advances in the use of energy concepts for seismic-resistant design have been achieved through analytical studies conducted on SDOFS. Because real structures, and particularly building structures, are MDOFS, there is an urgent need to conduct integrated analytical and experimental studies on the validity of applying the results obtained from analysis of SDOFS to MDOFS. To achieve this, the following recommendations are proposed: **(1)** To develop more efficient and reliable computer programs for the 3D, nonlinear dynamic analysis of multi-degree of freedom building structures; **(2)** to properly instrument whole multi-story building systems (soil-foundation-superstructure and nonstructural components) having different structural systems; **(3)** to conduct integrated and experimental investigation on the energy dissipation capacities of the different building structural components as well as their basic subassemblages when these components and assemblies are subjected to excitations reliably simulating the effects of the response of the building system to critical ground motions; **(4)** to use earthquake simulator facilities to perform integrated analytical and experimental studies on the 3D seismic performance of different types of building systems.

　　Development Needs.　　It is necessary that the results of the research be used to develop practical methods for applying the energy concepts, as well as the derived energy equations, to the design of buildings. To facilitate this application, it will also be necessary to develop efficient base isolation and base dissipation devices that can be used to control in a reliable way the E_I and E_{II} as well as the E_t.

Education Needs. It is time that the use of the energy concepts be introduced into the education of our engineering students. Furthermore, the practitioners (professional engineers) in regions of seismic risk should also be educated or at least exposed to the use of energy concepts. This information may be disseminated through short courses.

Implementation Needs. Researchers should work with professional engineers and code officials to develop practical methods of design based on energy concepts that can be introduced into the seismic codes.

ACKNOWLEDGEMENTS

The studies reported herein have been supported by grants from the National Science Foundation and CURFe Kajima. The authors would also like to acknowledge Dr. Eduardo Miranda, Research Engineer; Hatem Goucha, Research Assistant; and Katy Grether who edited and typed this paper.

REFERENCES

1. Bertero, V.V., State of the Art in Seismic-Resistant Construction of Structures. Proceedings, Third International Microzonation Conference, Seattle, WA, June 28-July 1, 1982, **2**, 767-808.

2. Bertero, V.V., Implications of Recent Earthquakes and Research on Earthquake-Resistant Design and Construction of Buildings. Report No. UCB/EERC-86/03, Earthquake Engineering Research Center, University of California, Berkeley, CA, March, 1986.

3. Bertero, V.V., Ductility Based Structural Design. Proceedings, Ninth World Conference on Earthquake Engineering, Tokyo-Kyoto, Japan, August, 1988, **8**, 673-686.

4. Bertero, V.V., Earthquake-Resistant Code Regulations of Recent Earthquakes and Research. Proceedings, International Symposium on Building Technology and Earthquake Hazard Mitigation, Kunming, China, March 25-28, 1991, 33 p.

5. Bertero, V.V., Strength and Deformation Capacities of Buildings Under Extreme Environments. In Structural Engineering and Structural Mechanics, A Volume Honoring Egor R. Popov, ed. Karl S. Pister, Prentice Hall, Inc., 1980, pp.188-237.

6. Uang, C-M., and Bertero, V.V., Implications of Recorded Ground Motions on Seismic Design of Structures. Report No. UCB/EERC 88/13, Earthquake Engineering Research Center, University of California, Berkeley, CA, Nov. 1988, 100 p.

7. Bertero, V.V., Structural Engineering Aspects of Seismic Zonation. Proceedings, Fourth International Conference on Seismic Zonation, Stanford University, Stanford, CA, August 26-29, 1991, 61 p.

8. Newmark, N.M. and Hall, W.J., Procedures and Criteria for Earthquake-Resistant Design. Building Science Series No. 46, Building Practices for Disaster Mitigation, National Bureau of Standards, Feb. 1973, pp.209-236.

9. Housner, G.W., Limit Design of Structures to Resist Earthquakes. Proceedings, First World Conference on Earthquake Engineering, Berkeley, CA, 1956, pp.5-1 to 5-13.

10. Berg, G.V. and Thomaides, S.S., Energy Consumption by Structures in Strong-Motion Earthquakes. Proceedings, Second World Conference on Earthquake Engineering, Tokyo, Japan, 1960, 2, 681-696.

11. Anderson, J.C. and Bertero, V.V., Seismic Behavior of Multi-Story Frames by Different Philosophies. Report no. UCB/EERC 69/11, Earthquake Engineering Research Center, University of California, Berkeley, CA, October 1969, 196 p.

12. Akiyama, H., Earthquake Resistant Limit-State Design for Buildings, University of Tokyo Press, 1985, 372 p.

13. Uang, C-M., and Bertero, V.V., Use of Energy as Design Criterion in Earthquake-Resistant Design. Report No. UCB/EERC-88/18, Earthquake Engineering Research Center, University of California, Berkeley, CA, No. 1988, 46 p.

14. Iwan, W.D., Estimating Inelastic Response Spectra from Elastic Spectra. Journal of Earthquake Engineering and Structural Dynamics, 1980, 8, 375-388.

15. Bertero, V.V. and Whittaker, A.S., Seismic Upgrading of Existing Buildings. 5as Jornadas Chilenas de Sismología è Ingeniería Antisísmica, Pontificia Universidad Católica de Chile, Agosto 1989, 1, 27-46.

SEISMIC DESIGN BASED ON DUCTILITY AND CUMULATIVE DAMAGE DEMANDS AND CAPACITIES

HELMUT KRAWINKLER and ALADDIN A. NASSAR
Department of Civil Engineering
Stanford University
Stanford, California 94305-4020, U.S.A.

ABSTRACT

This paper summarizes results of a continuing study on damage potential of ground motions and its implications for seismic design. A seismic design procedure that accounts explicitly for ductility and cumulative damage demands and capacities is proposed. The discussion focuses on the identification and determination of seismic demand parameters that are needed to implement the proposed design procedure. Emphasis is placed on strength, ductility, and energy demands. Results are presented for demands imposed by rock and stiff soil ground motions on single and multi degree of freedom systems. The objective of the paper is to demonstrate that ductility and cumulative damage considerations can and should be incorporated explicitly in the design process.

INTRODUCTION

Seismic design is an attempt to assure that strength and deformation capacities of structures exceed the demands imposed by severe earthquakes with an adequate margin of safety. This simple statement is difficult to implement because both demands and capacities are inherently uncertain and dependent on a great number of variables. A desirable long-range objective of research in earthquake engineering is to provide the basic knowledge needed to permit an explicit yet simple incorporation of relevant demand and capacity parameters in the design process. A demand parameter is defined here as a quantity that relates seismic input (ground motion) to structural response. Thus, it is a response quantity, obtained by filtering the ground motion through a linear or nonlinear structural filter. A simple example of a demand parameter is the acceleration response spectrum, which identifies the strength demand for elastic single degree of freedom (*SDOF*) systems. Considering that most structures behave inelastically in a major earthquake, it is evident that this parameter alone is insufficient to describe seismic demands. Relevant demand parameters include, but are not limited to, ductility demand, inelastic strength demand, and cumulative damage demands.

Capacities of elements and structures need to be described in terms of the same parameters as the demands in order to accommodate a design process in which capacities and demands can be compared directly. In this respect a clear distinction needs to be made between "brittle" elements and elements that tolerate inelastic deformations (ductile elements). In the former kind no reliance can be placed on ductility and the design process becomes an attempt to assure that the strength demands on these elements do not exceed the available strength capacities. This is usually accomplished by tuning the relative strength of ductile and brittle elements (e.g., the strong column - weak girder concept). The presence of ductile elements provides the opportunity to design structures for less strength capacity than the elastic strength demand imposed by ground motions by relying on the ductility capacity of these elements. The permissible amount of strength reduction depends on the ductility capacity, which in turn depends on the number, sequence, and magnitudes of the inelastic excursions (or cycles) to which the elements are subjected in an earthquake. This history dependence of ductility capacity, represented usually through cumulative damage models, presents one of the biggest challenges in improving seismic design procedures since it requires refined modeling that considers all important ground motion as well as structural response characteristics.

The work discussed in this paper addresses important issues in the context of seismic design for ductility demand and capacity, considering the effects of cumulative damage on the latter. A design procedure is postulated and the components of knowledge needed to implement this procedure are identified. The issue of cumulative damage is briefly discussed, quantitative information is presented on important seismic demand parameters for *SDOF* systems, and selected data are presented for multi-degree of freedom (*MDOF*) systems that can be viewed as conceptual models of real multi-story buildings.

POSTULATED SEISMIC DESIGN PROCEDURE

The objective is to develop a design approach that permits better tuning of the design to the ductility capacities of different structural systems and the elements that control seismic behavior. Such an approach has to be simple to be adopted by design engineers and transparent to the design process to permit the designer to explicitly consider demands versus capacities. The approach must be applicable equally to the limit states of serviceability and safety against collapse (i.e, a dual level design approach). Both limit states can be described by damage control, with the serviceability limit state defined by drift limitations and small cumulative damage, and the safety limit state defined by an adequate margin of safety against the cumulative damage approaching a limit value associated with collapse. The discussion presented in this paper is not concerned with the issue of serviceability. It focuses on design for safety against collapse during severe earthquakes.

In the design for safety against collapse it is postulated that element behavior can be described by cumulative damage models of the type summarized in the next section. Since these damage models are too complex to be incorporated directly into the design process, it is suggested to use these models together with statistical information on seismic demand parameters and experimental and analytical data on structural performance parameters to transform element cumulative damage capacity into element ductility capacity (ductility capacity weighted with respect to anticipated cumulative damage demands such as hysteretic energy dissipation). Thus, the <u>ductility capacity</u> of the critical structural elements becomes the starting point for seismic design. This capacity will depend on the types of elements used in the structural system, but once the elements have been chosen the ductility capacity is assumed to be a known quantity. The strengths of elements and the structure become now dependent quantities which need to be derived from the criterion that the ductility demands should not exceed the known ductility capacities.

In order to derive structure strength requirements, the element ductility capacities have to be transformed into story ductility capacities (sometimes a simple geometric transformation and sometimes an elaborate process), which are then used to derive "inelastic strength demands" for design (discussed later). The so derived strength demands identify the required ultimate strength of the structure. Recognizing that the design profession prefers to perform elastic rather than plastic design, the structure strength level may be transformed to the member strength level in order to perform conventional elastic strength design (by estimating the ratio of the strength of the structure to the strength level associated with the end of elastic response, shown as E_g and E_l, respectively, in Figure 1). Pilot studies have shown that for regular structures this transformation is usually not difficult but may require an iteration [1]. After this preliminary design an important step is design verification through a *nonlinear static incremental load analysis* (using a rational static load pattern in a "push-over" loading) to verify that the required structure strength (E_g) is achieved and that "brittle" elements are not overloaded (ductility demand < 1.0).

Figure 1 illustrates the step-by-step implementation of the proposed design approach. As a basis the implementation requires a model to weigh ductility capacity for anticipated cumulative damage effects. In the illustration equal normalized hysteretic energy dissipation (see next section) is used as a criterion for weighing ductility capacity. Using the fundamental period of the structure, T, and its weighted ductility capacity, μ, the strength reduction factor, R, can be evaluated from R-μ-T relationships discussed later, assuming the structure can be modeled as an *SDOF* system. This strength reduction factor is used to scale the elastic strength demand spectrum (i.e., the acceleration response spectrum) to obtain inelastic strength demands. System dependent modification factors are then applied to the *SDOF* inelastic strength demands to account for higher mode effects in *MDOF* systems. This step identifies the structure strength demand, E_g, which defines the strength capacity required in order to limit the ductility demands on the structural elements to the target ductility capacities. The local strength demand (associated with the end of elastic response), E_l, is then estimated from the structure strength demand, the structure is designed employing conventional elastic strength design, and a nonlinear incremental load analysis is carried out to verify required structure strength.

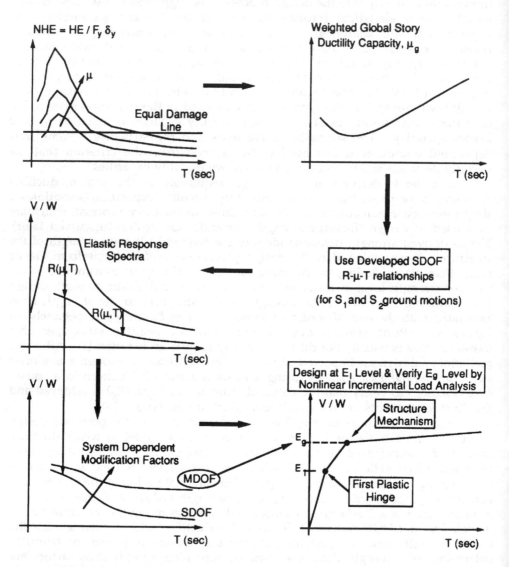

FIGURE 1. Implementation of postulated seismic design procedure

Clearly, there are many issues in this design approach that have not been addressed and that may complicate the process considerably. But the approach has been shown to work in simple examples [1], and deserves further study to explore its potential. The following list itemizes the basic information needed to implement this approach.

1. Experimental and analytical information on cumulative damage models for structural elements.
2. Statistical data on anticipated cumulative damage demands needed to weigh ductility capacities.
3. Statistical data on inelastic strength demands for prescribed ductility capacities, using *SDOF* systems.
4. Statistical data on multi-mode effects on the inelastic strength demands derived from *SDOF* systems.

The following sections provide discussions on specific aspects of these four items. More detailed information is presented in Reference [2].

EXAMPLES OF CUMULATIVE DAMAGE MODELS

It is well established from experimental work and analytical studies that strength and stiffness properties of elements and structures deteriorate during cyclic loading. Materials, and therefore elements and structures, have a memory of past loading history, and the current deformation state depends on the cumulative damage effect of all past states. In concept every excursion causes damage, and damage accumulates as the number of excursions increases. The damage caused by elastic excursions is usually small and negligible in the context of seismic behavior. Thus, only inelastic excursions need to be considered, and from those the large ones cause significantly more damage than smaller ones (however, smaller excursions are much more frequent).

Many cumulative damage models have been proposed in the literature, each one of them with specific materials, elements, and failure modes in mind. None of the proposed models is universally applicable. A comprehensive summary of widely used model is provided in [3]. Only two of these models are summarized here, the first one developed specifically for elements of reinforced concrete structures, and the second one developed primarily for elements of steel structures.

The damage index proposed by Park and Ang [4] for reinforced concrete elements is expressed as a linear combination of the normalized maximum deformation and the normalized hysteretic energy as follows:

$$D = \frac{\delta_m}{\delta_u} + \frac{\beta}{F_y \, \delta_u} \int dE$$

(1)

in which
D = damage index ($D > 1$ indicates excessive damage or collapse)
δ_m = maximum deformation under earthquake
δ_u = ultimate deformation capacity under static loading
F_y = calculated yield strength
dE = incremental hysteretic energy
β = parameter accounting for cyclic loading effect

Park and Ang [4] tested this model on 403 specimens and found that the damage capacity D is reasonably lognormal distributed but that the data show considerable scatter (c.o.v. = 0.54), which is to be expected for reinforced concrete elements. This model is simple to apply and has been used widely for damage evaluation of reinforced concrete structures.

The cumulative damage model proposed by Krawinkler and Zohrei [5] takes on the following form:

$$D = C \sum_{i=1}^{N} (\Delta \delta_{pi} / \delta_y)^c$$

(2)

in which D = damage index ($D > 1$ indicates excessive damage or collapse)
$\Delta \delta_{pi}$ = plastic deformation range of excursion i (see Figure 2)
δ_y = yield deformation
N = number of inelastic excursions experienced in earthquake
C, c = structural performance parameters

This model was tested and found to give very good results for several failure modes in elements of steel structures [5]. In these tests the exponent c was found to be in the order of 1.5 to 2.0, whereas the coefficient C varies widely and depends strongly on the performance characteristics of the structural element. The model has not been tested on reinforced concrete elements.

The two models appear to be very different but, in fact, they are rather similar under specific conditions. Both contain two structural performance parameters, δ_u and β in the Park/Ang model and C and c in the Krawinkler/Zohrei model. Both contain, explicitly or implicitly, normalized hysteretic energy dissipation as the primary cumulative damage parameter. This is evident in the Park/Ang model in which the hysteretic energy dissipated in each cycle is normalized by the product $F_y \delta_u$. In the Krawinkler/Zohrei model hysteretic energy dissipation is contained in the term $\Sigma(\Delta \delta_{pi}/\delta_y)$, which for elastic-plastic structural systems is equal to the hysteretic energy dissipation normalized by $F_y \delta_y$. It can be shown that this relationship is almost exact also for bilinear strain hardening systems of the type illustrated in Figure 2.

Thus, the hysteretic energy dissipation, HE, or its normalized value, $NHE = HE/F_y \delta_y$, is believed to be the most important cumulative damage parameter. It is evident that the hysteretic energy <u>demand</u> depends strongly on the strong motion duration, frequency content of ground motions, and the period and yield level of the structure, since they all affect the number and magnitudes of inelastic excursions, which in turn determine the cumulative damage experienced by a structure. Moreover, hysteretic energy dissipation is only one of the terms involved in the energy equilibrium of a structure, and the hysteretic energy demand imposed by a ground motion depends also on the other energy terms (i.e., damping energy DE, kinetic energy KE, and recoverable strain energy RSE) that make up the input energy, IE, imparted to the structure by a ground motion.

This brief discussion on cumulative damage modeling was intended to show that damage and energy demands are closely related. Evaluation of energy demands is important in seismic design for two reasons. For one, input energy demand spectra, which include all energy components ($RSE, KE, DE,$ and HE), give a clear picture of the cumulative damage potential of ground motions, much more so than elastic response spectra. Secondly, hysteretic energy demand spectra, which form an important part of the total input energy demand spectra, serve to provide the information necessary to modify ductility capacities in accordance with appropriate cumulative damage models of the type discussed in this section. Thus, there are good reasons to evaluate energy demand spectra in addition to other demand spectra discussed in the next section.

SEISMIC DEMAND PARAMETERS NEEDED FOR DESIGN

Seismic demands represent the requirements imposed by ground motions on relevant structural performance parameters. In a local domain this could be the demand on axial load of a column or the rotation of a plastic hinge in a beam, etc. Thus, the localized demands depend on many local and global response characteristics of structures, which cannot be considered in a study that is concerned with a global evaluation of seismic demands. In this study only *SDOF* systems and simplified *MDOF* systems are used as structural models. Assuming that these models have a reasonably well defined yield strength, the following basic seismic demand parameters play an important role in implementing the postulated design procedure. Some of the terms used in these definitions are illustrated in Figure 2.

Elastic Strength Demand, $F_{y,e}$. This parameter defines the yield strength required of the structural system in order to respond elastically to a ground motion. For *SDOF* systems the elastic response spectra provide the needed information on this parameter.

Ductility Demand, μ. This parameter is defined as the ratio of maximum deformation over yield deformation for a system with a yield strength smaller than the elastic strength demand $F_{y,e}$.

Inelastic Strength Demand, $F_y(\mu)$. This parameter defines the yield strength required of an inelastic system in order to limit the ductility demand to a value of μ.

Strength Reduction Factor, $R_y(\mu)$. This parameter defines the reduction in elastic strength that will result in a ductility demand of μ. Thus, $R_y(\mu) = F_{y,e}/F_y(\mu)$. This parameter is often denoted as R.

Energy and Cumulative Damage Demands. From the cumulative damage parameters discussed in the previous section only the following two are discussed here:

Hysteretic Energy, *HE*:   The energy dissipated in the structure through inelastic deformation

Total Dissipated Energy, *TDE*: $TDE = HE + DE$ (*TDE* is usually equal to the maximum input energy *IE* except for short period structures and structures with very large velocity pulses).

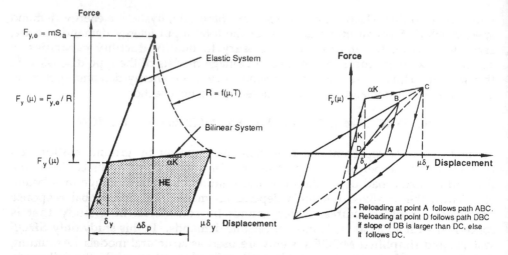

FIGURE 2. Basic seismic demand parameters   FIG. 3. Stiffness degrading model

The list of seismic demand parameters enumerated here is by no means complete.   But for conceptual studies much can be learned from these parameters.   In the following section these parameters are evaluated for closely spaced periods of two types of *SDOF* systems in order to permit a representation in terms of spectra, using a period range from 0.1 sec. to 4.0 sec. In the subsequent section the strength and ductility demands are evaluated for three types of *MDOF* systems, using six discrete periods covering a range from 0.22 to 2.05 seconds.

STATISTICAL DATA ON *SDOF* SEISMIC DEMANDS FOR ROCK AND STIFF SOIL GROUND MOTIONS

The results discussed here are derived from a statistical study that uses 15 Western US ground motion records from earthquakes ranging in magnitude from 5.7 to 7.7. All records are from sites corresponding to soil type S_1 (rock or stiff soils). Time history analysis was performed with each record, using bilinear (see Figure 2) and stiffness degrading (see Figure 3) *SDOF* systems in which the yield levels are adjusted so that discrete predefined target ductility ratios of 2, 3, 4, 5, 6, and 8 are achieved. Damping of 5% of critical was used in all analyses and strain hardening of $\alpha = 0$, 2%, and 10% was investigated.

Since the problem of scaling records to a common severity level is an unresolved issue, all results shown here are presented in a form that makes scaling unnecessary. This is accomplished by computing for each record the demand parameters for constant ductility ratios and normalizing the demand parameters by quantities that render the results dimensionless. The normalized parameters are then evaluated statistically. Only sample mean values are presented here.

Typical mean spectra of normalized hysteretic energy, $NHE = HE/F_y\delta_y$, for bilinear *SDOF* systems are shown in Figure 4. The graphs show the significant effect of system period on this parameter, particularly for higher ductility ratios.

Thus, if constant *NHE* were a measure of equal damage, it would be prudent to limit the ductility capacity for short period structures to significantly lower values than for long period structures. What is not shown in these mean spectra is the effect of strong motion duration on *NHE*. It is recognized that this effect is strong, but no success can be reported in our attempts to correlate *NHE* and strong motion duration, even when employing several of the presently used definitions of strong motion duration.

The effect of different hysteresis models (stiffness degrading versus bilinear) on *NHE* is illustrated in Figure 5. In general, and particularly for short period systems, the stiffness degrading model needs to dissipate more hysteretic energy than the bilinear model. The reason is simply that the stiffness degrading model executes many more small inelastic excursions than the bilinear model in which many excursions remain in the elastic range.

The contribution of hysteretic energy dissipation to the total dissipated energy *TDE* for bilinear systems is illustrated in Figure 6. These graphs are valid only for systems with 5% damping. It can be seen that the ratio *HE/TDE* is not very sensitive to the ductility ratio except for low ductilities. It was found that this ratio is very stable for all records used in this study. Thus, the presented data can be used to evaluate the effectiveness of viscous damping compared to hysteretic energy dissipation in dissipating the energy imparted to a structure. As Figure 7 shows, in stiffness degrading systems a much larger portion of *TDE* is dissipated through inelastic deformations (hysteretic energy) than in bilinear systems, indicating that viscous damping is much less effective in stiffness degrading systems.

In the context of the postulated design procedure, the energy demands illustrated here provide information to be used to modify ductility capacities by means of cumulative damage models. In the design process, the need exists then to derive the strength required so that the ductility demands are limited to the target ductility capacities. These strength demands can be represented by inelastic strength demand spectra or, in dimensionless form, in terms of the strength reduction factor *R*, which is the ratio of elastic strength demand, $F_{y,e}$, over inelastic strength demand for a specified target ductility ratio, $F_y(\mu)$. A two-step nonlinear regression analysis was performed on the *R*-factors, first regressing *R* versus μ for constant periods *T*, and then evaluating the effect of period in a second step. For reasons discussed in Reference [2], the following form of *R*-μ-*T* relationship was employed:

$$R = \left\{ c\,(\mu - 1) + 1 \right\}^{1/c} \quad \text{where} \quad c(T, \alpha) = \frac{T^a}{1 + T^a} + \frac{b}{T} \tag{3}$$

For different strain hardening ratios α the following values were obtained for the two regression parameters *a* and *b*:

$$
\begin{array}{lll}
\text{for } \alpha = 0\%: & a = 1.00 & b = 0.42 \\
\text{for } \alpha = 2\%: & a = 1.00 & b = 0.37 \\
\text{for } \alpha = 10\%: & a = 0.80 & b = 0.29 \\
\end{array}
$$

The regression curves for $\mu = 2, 3, 4, 5, 6,$ and 8 for bilinear systems with 10% strain hardening are shown in Figure 8 together with the mean values of

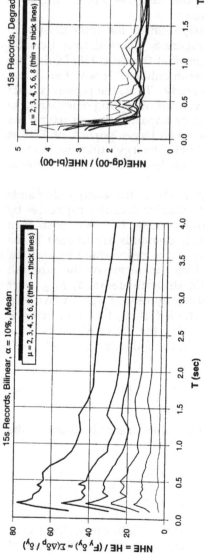

FIGURE 4. Mean spectra of normalized hysteretic energy, *NHE* (bilinear, α = 10%)

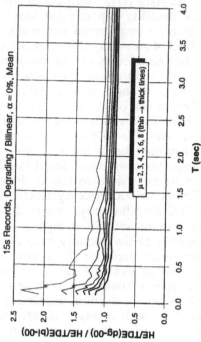

FIGURE 5. Ratio of *NHE* for stiffness degrading to bilinear systems (α = 0)

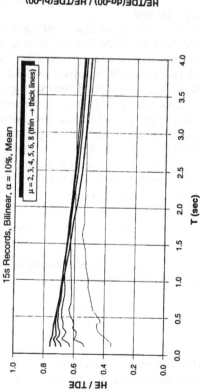

FIGURE 6. Contribution of *HE* to total dissipated energy, *TDE* (bilinear, α = 10%)

FIGURE 7. Ratio of *HE/TDE* for stiffness degrading to bilinear systems (α = 0)

the data points on which the regression was based. It is evident that the R-μ-T relationships are nonlinear, particularly in the short period range. For all ductility ratios the R-factors approach 1.0 as T approaches zero, and they approach μ as T approaches infinity.

Relationships of this type together with mean or smoothened elastic response spectra can be employed in many cases to evaluate the inelastic strength demands. This can be done with confidence for S_1 soil types, on which these relationships are based, and probably also for S_2 soil types since the average R-factors were found to be insensitive to relatively small variations in average response spectra shapes. However, these R-μ-T relationships cannot be applied to motions in soft soils which contain a signature of the site soil column. If we use these R-μ-T relationships to derive inelastic strength demand spectra from the ATC S_1 ground motion spectrum, the results shown in Figure 9 are obtained. To no surprise, the inelastic strength demands are anything but constant for periods below 0.5 sec., the range in which the smoothened elastic response spectrum has a plateau.

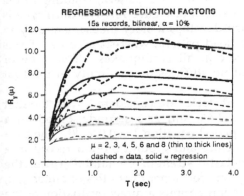

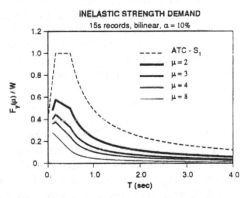

FIG. 8. Strength reduction factor $R_y(\mu)$ for motions in S_1 soil type

FIG. 9. Inelastic strength demand spectra based on ATC S_1 spectrum

The R-factors presented in Figure 8 can be used with good confidence also for stiffness degrading systems of the type shown in Figure 3. From the statistical study it was found that this type of stiffness degradation has only a small effect on the strength demands for all periods and ductility ratios. The same cannot be said about the effect of stiffness degradation on energy demands, as is evident from Figures 5 and 7.

EFFECTS OF HIGHER MODES ON INELASTIC STRENGTH DEMANDS

The previous section provided information on seismic demands for inelastic $SDOF$ systems. This information is relevant as baseline data but needs to be modified to become of direct use for design of real structures, which mostly are multi-degree-of-freedom ($MDOF$) systems affected by several modes. For inelastic $MDOF$ systems, modal superposition cannot be applied with any degree of confidence and different techniques have to be employed in order to predict strength or ductility demands that can be used for design.

The research summarized here is intended to provide some of the answers needed to assess strength demands for inelastic $MDOF$ systems. The focus is on a statistical evaluation of systems that are regular from the viewpoint of elastic dynamic behavior. Thus, closely spaced modes and torsional effects are neglected and structures are modeled two-dimensionally. For convenience in computer analysis, all structures are modeled as single bay frames even though they are intended to represent generic structures with three distinctly different types of behavior patterns.

The three types of structures are illustrated in Figure 10. The first type is designated as "beam hinge model" (strong column - weak beam model), from here on referred to as BH model, and represents structures that develop under the 1988 UBC seismic load pattern a complete mechanism with plastic hinges in all beams forming simultaneously as shown in Figure 10. The second type is designated as "column hinge model" (weak column - strong beam model), or CH model. It represents structures whose relative column strengths are tuned in a manner such that all columns simultaneously develop plastic hinges under lateral loads corresponding to the 1988 UBC seismic load pattern, resulting in the "collapse" mechanism shown in the second sketch of Figure 10. The third type is a "weak story model," or WS model, which develops a story mechanism only in the first story under the 1988 UBC seismic load pattern, whereas all other stories are of sufficient strength to remain elastic in all earthquakes. This type of structure has a strength discontinuity but no stiffness discontinuity in the first story.

Structures with 2, 5, 10, 20, 30, and 40 stories are considered, with the first mode periods being 0.22, 0.43, 0.73, 1.22, 1.65, and 2.05 seconds, based on a constant story height of 12 ft and the code period equation $T = 0.02h_n^{3/4}$. The base shear strength, V_y, is varied for each structure and ground motion record in a manner so that it is identical to the inelastic strength demand $F_y(\mu)$ of the corresponding first mode period $SDOF$ system for target ductilities of either 1, 2, 3, 4, 5, 6, or 8. Applying this strength criterion permits a direct evaluation of the differences between $SDOF$ and $MDOF$ responses for each ground motion.

A total of 5,670 nonlinear time history analyses were performed, using the 15 S_1 ground motion records, 3 types of structures (BH, CH, and WS), 6 different numbers of stories, 7 different yield levels (corresponding to $SDOF$ yield strengths for $\mu = 1, 2, 3, 4, 5, 6$, and 8), and 3 strain hardening ratios ($\alpha = 0, 2\%$, and 10%). Response parameters obtained from the 15 records were evaluated statistically using sample means and variations. The results of this study are discussed in detail in [2], and only a few pertinent data are summarized below.

Figure 11 shows typical results of mean values of story ductility ratios for the three types of structures. The graphs apply for structures whose base shear strength is equal to the $SDOF$ inelastic strength demand for a target ductility ratio of 8. The resulting ductility ratios may be unrealistic, but this example is chosen to clearly illustrate consistent trends and patterns. It is observed that the story ductility demands for the $MDOF$ systems are largest in the bottom story (this was found to be true for most cases but not necessarily always for lower target ductility ratios) and in this story are larger than the $SDOF$ target ductility ratio of 8 because of higher mode effects. The increase above the target ductility ratio is smallest for the BH structures and by far the highest for the WS structures. This observation was found to hold true regardless of the number of stories, the target ductility ratio, and the strain hardening ratio, which clearly illustrates the importance of higher mode effects and of the type of "failure" mechanism inherent in the structural system.

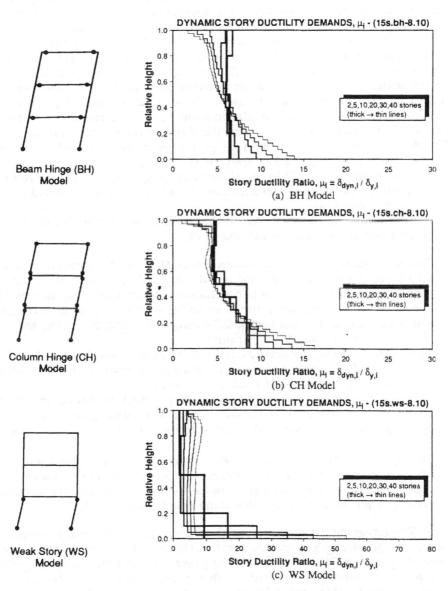

Beam Hinge (BH)
Model

(a) BH Model

Column Hinge (CH)
Model

(b) CH Model

Weak Story (WS)
Model

(c) WS Model

FIGURE 10. Types of models used in this MDOF study

FIGURE 11. *MDOF* story ductility demands for base shear strength corresponding to *SDOF* target ductility ratio of 8 ($\alpha = 10\%$)

In the postulated design procedure the objective is to limit the story ductility ratios to predetermined target values. The results illustrated in Figure 11 show clearly that the base shear strength obtained from the corresponding SDOF system is insufficient to achieve this objective. Thus, the inelastic strength demands obtained for SDOF systems must be modified to to be applicable to MDOF structures. The modification depends on the number of stories (first mode period), the target ductility ratio, the strain hardening ratio, and the type of "failure" mechanism in the structure. For the three types of structure investigated here, data of the kind presented in Figure 11 can be utilized to derive the necessary modifications [2]. Examples of derived modification factors are presented in Figure 12 for target ductility ratios of 4 and 8. The modification factors define the required increase in base shear strength V_y of the MDOF structure over the inelastic strength demand F_y of the corresponding first mode period SDOF system in order to limit the ductility ratio to the same target value. The dashed curves shown in the four graphs represent the modification factors implied by the widely used procedure of raising the $1/T$ tail of the ground motion spectrum to $1/T^{2/3}$ in the elastic design spectrum. This procedure was first introduced in the ATC 3-06 document [6] and is presently adopted in the U.S. Uniform Building Code.

The following observations can be made from Figure 12 and similar but more comprehensive graphs presented in [2].

- The required strength modifications are smallest for BH structures. For these structures the modifications are mostly in good agreement with the ATC-3 modification provided there is considerable strain hardening ($\alpha = 10\%$). For short period BH structures the base shear strength demand is consistently lower than the corresponding SDOF strength demand, indicating that MDOF effects are not important in this range.
- The MDOF strength demands for CH structures are higher than for BH structures. The required increase in strength compared to BH structures is about the same regardless of fundamental period.
- The modification factor increases with target ductility ratios and decreases with strain hardening. Systems without strain hardening ($\alpha = 0\%$) drift more, and larger strength is required in order to limit the drift to a prescribed target ductility ratio.
- The figure clearly illustrates that WS structures, i.e., structures with a weak first story, are indeed a great problem. Such structures require strength capacities that may be more than twice those required for BH structures in order to limit the story drift to the same target ductility ratio.

The foregoing discussion focused on a procedure that can be employed to derive design strength demands for MDOF systems from inelastic strength demand spectra of SDOF systems. The presented numerical results apply only within the constraints identified in this section and cannot be generalized without a much more comprehensive parametric study. The parameters that need to be considered include the frequency content of the ground motions (which may be greatly affected by local site conditions), the hysteretic characteristics of the structural models (stiffness degradation, strength deterioration, etc.), and the dynamic characteristics of the MDOF structures

37

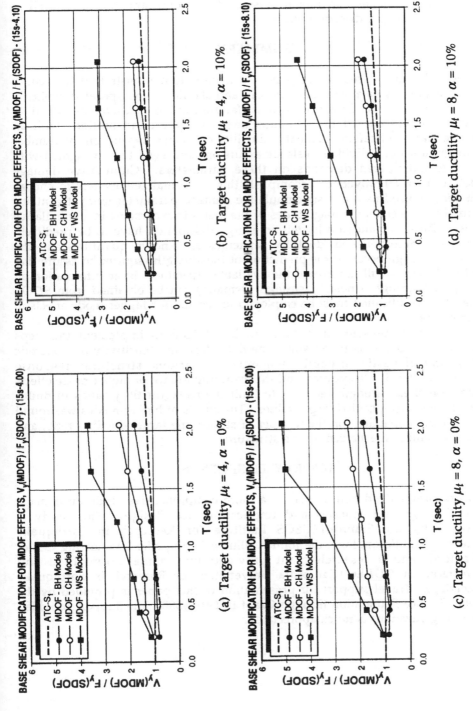

(a) Target ductility $\mu_t = 4$, $\alpha = 0\%$

(b) Target ductility $\mu_t = 4$, $\alpha = 10\%$

(c) Target ductility $\mu_t = 8$, $\alpha = 0\%$

(d) Target ductility $\mu_t = 8$, $\alpha = 10\%$

FIGURE 12. Modifications in base shear strength required to account for *MDOF* effects (periods, mode shapes and modal masses of all important modes, as well as stiffness and strength discontinuities).

SUMMARY

The research summarized in this paper is intended to demonstrate that ductility and cumulative damage consideration can and should be incorporated explicitly in the design process. Protection against failure implies that available ductility capacities should exceed the demands imposed by ground motions with an adequate margin of safety. Available ductility capacities depend on the number and magnitudes of individual inelastic excursions and need to be weighted with respect to anticipated demands on these parameters. Cumulative damage models can be employed to accomplish this. Normalized hysteretic energy dissipation is used as the basic cumulative damage parameter since it contains the number as well as the magnitudes of the inelastic excursions in a cumulative manner. Thus, demands on hysteretic energy dissipation have to be predicted. Once this is accomplished, ductility capacities are known quantities and the objective of design becomes the prediction of the strength required to assure that ductility demands will not exceed the available capacities. Basic information on the required strength (inelastic strength demand) can be obtained from *SDOF* studies, but modification must be employed to account for higher mode effects in real structures.

The paper presents data that can be utilized to implement the steps outlined in the previous paragraph. The data show the sensitivity of hysteretic energy and inelastic strength demands to various structural response characteristics for *SDOF* systems, and the great importance of higher mode effects on the base shear strength required to limit the story ductility ratios in multi-story structures to specified target values. The effects of higher modes was found to be strongly dependent on the number of stories, the target ductility ratio, and the type of failure mechanism in the structure.

ACKNOWLEDGEMENTS

The work summarized here is part of a more comprehensive study on damage potential of ground motions and implications for design. This study is supported by the Stanford/USGS Institute for Research in Earthquake Engineering and Seismology, the John A. Blume Earthquake Engineering Center at Stanford, and a research grant provided by Kajima Corporation and administered by CUREe (California Universities for Research in Earthquake Engineering). Travel support for presentation of this paper is provided by the National Science Foundation through Grant INT-9114580. The support provided by these organizations is much appreciated.

REFERENCES

1. Osteraas, J.D., and Krawinkler, H., Strength and ductility considerations in seismic design. John A. Blume Earthquake Engineering Center Report No. 90, Department of Civil Engineering, Stanford University, August 1990.

2. Nassar, A.A., and Krawinkler, H., Seismic demands for SDOF and MDOF systems. John A. Blume Earthquake Engineering Center Report No. 95, Department of Civil Engineering, Stanford University, June 1991.

3. Chung, Y.S., Meyer, C., and Shinozuka, M., Seismic damage assessment of reinforced concrete members. Report NCEER-87-0022, National Center for Earthquake Engineering Research, State University of New York at Buffalo, October 1987.

4. Park, Y.-J., and Ang, A.H.-S., Mechanistic seismic damage model for reinforced concrete. Journal of Structural Engineering, ASCE, Vol.111, No. 4, April 1985, pp. 722-739.

5. Krawinkler, H., and Zohrei, M., Cumulative damage in steel structures subjected to earthquake ground motions. Journal on Computers and Structures, Vol. 16, No. 1-4, 1983, pp. 531-541.

6. Tentative provisions for the development of seismic regulations for buildings. Applied Technology Council Report ATC 3-06, June 1978.

ON ENERGY DEMAND AND SUPPLY IN SDOF SYSTEMS

PETER FAJFAR, TOMAŽ VIDIC, MATEJ FISCHINGER
Department of Civil Engineering
University of Ljubljana
Jamova 2, 61000 Ljubljana, Slovenia

ABSTRACT

In the paper a brief overview of research results on the seismic energy demand and supply of SDOF systems, obtained in recent years at the University of Ljubljana, is presented. The first part of the paper deals with energy demand. The results of parametric studies on input energy, on the hysteretic to input energy ratio, and on the relation between hysteretic energy and maximum displacement are presented, and simple formulae, which can be used in design, are proposed. Energy dissipation capacity (supply) is defined in terms of equivalent ductility factors, which represent reduced values of ductility factors based on maximum displacement under monotonic loading.

INTRODUCTION

Aseismic design philosophy for ordinary building structures relies strongly on energy dissipation through large inelastic deformations. Structures are considered to have adequate seismic resistance if their limit capacity exceeds seismic demand in the case of severe earthquakes. Maximum relative displacement (or displacement ductility defined as the ratio of the maximum to the yield displacement) is the structural response parameter most widely employed for evaluating the inelastic performance of structures. Displacement ductility is also the parameter explicitly or implicitly used in most common design procedures (e.g. the Newmark-Hall method for constructing design spectra [1]). It has been widely recognized, however, that the level of structural damage due to earthquakes does not depend on maximum displacement only and that the cumulative damage resulting from numerous inelastic cycles should also be taken into account. The energy input E_I to a structure subjected to strong ground motion is dissipated in part by inelastic deformations (hysteretic energy E_H) and, in part, by viscous damping, which represents miscellaneous damping effects other than inelastic deformation (damping energy). Dissipated hysteretic energy is the structural response parameter which is supposed to be closely correlated to cumulative damage.

In the paper both seismic energy demand and seismic energy dissipation capacity (supply) of single-degree-of-freedom (SDOF) systems are discussed. A brief overview of research results obtained in recent years at the University of Ljubljana, and partially published or submitted for publication elsewhere [2-7] is presented.

The paper is restricted to SDOF systems. However, inelastic spectra have been widely used in practice for the analysis of inelastic MDOF systems. Methods for the inelastic analysis of MDOF systems based on equivalent SDOF systems have been developed [9,10]. Akiyama [11] has proposed a design method for both SDOF and MDOF systems on the basis of the energy concept. He has shown that there is no essential difference between SDOF and MDOF systems with regard to the total energy input per unit mass. Based on the above discussion, it can be concluded that the results obtained using SDOF systems, and reported in this paper, can be applied to many MDOF systems.

DATA FOR PARAMETRIC STUDIES ON SEISMIC ENERGY DEMAND

In order to identify the influence of the most important strong motion and structural parameters on the earthquake response of structures, parametric studies of SDOF systems were carried out. In these studies, input ground motion, as well as the initial stiffness (expressed by the natural period), strength, hysteretic behaviour and damping of SDOF structural systems, was varied. Seismic demand, expressed in terms of input energy per unit mass E_I/m, the hysteretic to input energy ratio E_H/E_I, and the parameter γ, which represents the relation between hysteretic energy and maximum displacement, were studied and presented as functions of the natural period T. The amount of dissipated hysteretic energy was determined at a time equal to that at the end of the ground motion plus two initial periods of the system.

Earthquake records

The influence of input motion has been studied using six different groups of records in order to take into account ground motions of basically differing types. The standard records from California and the records from Montenegro, Yugoslavia, 1979, were chosen as being representative for "standard" ground motion. The main characteristic of the Friuli, Italy, 1976, and Banja Luka, Yugoslavia, 1981, records is the short duration of the strong ground motion. The predominant periods of these records are short and fairly narrow-banded. The 1985 Mexico City records show ground motions of long duration, with long predominant periods. The duration of the 1985 Chile records is long, but their predominant periods are shorter than those of the "standard" records. A total of 48 horizontal components of records obtained at 24 different stations have been used. In a part of the study on input energy, additional records from Japan, Nahanni (Canada) and Bucharest (Romania) were used. The Japanese records were divided into two groups. 12 records from 6 stations (Japan 1) represent a collection without any specific common characteristic, whereas the main characteristic of 4 records (Japan 2) are long duration and a very short predominant period. In this respect these records are similar to records from Chile. The peak ground acceleration and velocity of these records , however, is much lower than in the case of the Chile records. The characteristics of the ground motion during the Nahanni, Canada, 1985 earthquake (the main event on December 23) are high peak accelerations, short predominant periods and short duration. Due to a combination of several shocks in the records, it is difficult to determine realistic values of strong motion duration by standard methods. The most important feature of the record obtained during the Vrancea earthquake, Rumania, 1977, in Bucharest (NS component), is a single large acceleration pulse connected with high peak ground velocity. The total number of additional records is 24. A list of all records is given in Table 1 (two horizontal components have been used in all cases). A detailed description of the records of the first five groups can be found in earlier papers by the authors [2,3], whereas all the records are more precisely defined in [7].

TABLE 1
List of records used in the parametric studies

Group	Stations
California	El Centro 1934, El Centro 1940, Olympia, Taft, Castaic
Friuli 1976	Tolmezzo, Forgaria, San Rocco
Montenegro 1979	Petrovac, Ulcinj (Albatros and Olympia), Bar, Hercegnovi
Banja Luka 1981	IMS, BK-2, BK-9
Mexico City 1985	SCT, Abastos, C. University, Viveros
Chile 1985	Llollco, Vina del Mar, Valparaiso (El Almendral and UTFSM)
Japan 1	Kochi (Apr. 1, 1968), Aomori (May 16, 1968), Murokan (May 16, 1968), Hososhima (July 26, 1970), Ofunato (June 12, 1978), Shiogama (June 12, 1978)
Japan 2	Miyako (May 16, 1968), Miyako (June 12, 1968)
Nahanni 1985	S1, S2, S3
Rumania 1977	Bucharest - INCERC

Usually, a distinctive single predominant period does not exist. A generally accepted definition of the predominant period is also lacking. The authors have applied the formula proposed by Heidebrecht [11] for the limit period between the short- and medium-period range

$$T_1 = 4.3 \frac{v_g}{a_g} \tag{1}$$

where v_g and a_g are, respectively, the peak ground velocity and acceleration.

Structural systems
The following structural parameters were varied in the studies: initial stiffness (period), strength, hysteresis and damping. The period range from 0.1 to 2.5 s was considered. The

value of the strength parameter η, which is defined as the yield resistance F_y divided by the mass of the system and by the peak ground acceleration a_g

$$\eta = \frac{F_y}{m\,a_g} \tag{2}$$

was varied from 5 (simulating elastic behaviour) to 0.2. In some cases constant ductility was assumed throughhout the whole period region. Ductility factors equal to 2, 4, and 6 were chosen. The relation between the strength parameters η and the ductility factors μ are shown in Fig. 6d. The η - T curves represent normalized inelastic design spectra based on constant ductility.

Damping can be assumed to be proportional to either mass or stiffness. In linear analysis both approaches yield the same results. In nonlinear analysis, however, stiffness degrades with damage. Consequently, damping related to instantaneous stiffness tends to decrease, whereas damping related to mass tends to increase with degrading of stiffness. How damping is mathematically modeled may have an important influence on some of the response parameters, e.g. on maximum displacements and dissipated hysteretic energy. Unfortunately, in the literature it is usually not clearly stated what type of damping was assumed. It is believed that mass-proportional damping is more often used. According to Otani [12], however, it is not realistic to expect mass-proportional damping in an actual reinforced concrete structure. In our study both mass-proportional viscous damping and viscous damping proportional to instantaneous stiffness (2 and 5 per cent) were used.

Altogether, eight hysteresis models were investigated (Fig. 1). Six of them simulate predominantly flexural behaviour: the elastoplastic, the bilinear, the Q-model and three variants of Takeda's model with a trilinear envelope (they differ according to unloading stiffness and envelope shape). Shear behaviour has been simulated by two variants of the shear-slip model. The results of the parametric study of different structural parameters are presented as mean values from the 20 records (the U.S.A. and Montenegro groups) representing "standard" ground motion. The average predominant period for these records is roughly $T_1 = 0.5$ s. A system with $\eta = 0.6$, Q-hysteresis and 5 per cent damping proportional to instantaneous stiffness was chosen as the basic "average inelastic system", which is supposed to represent an average reinforced concrete structure with a natural period in the medium-period region, designed according to the codes. (In the parametric study on input energy, which was carried out in the first phase of the parametric study, exceptionally 5 per cent of mass-proportional damping was assumed in the "average inelastic system".) In this system, the parameters were varied one by one. In addition, some other combinations of parameters were also investigated.

INPUT ENERGY

Uang and Bertero [13] have called attention to the difference between "absolute" and "relative" energy formulations, which is important in the very short and very long period ranges. In this study, the "relative" energy formulation has been used. Relative input energy

45

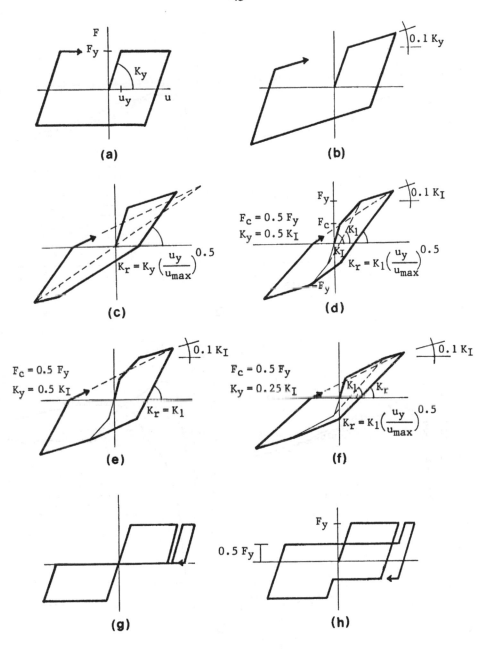

Figure 1. Hysteresis models
(a) elastoplastic, (b) bilinear, (c) Q-model, (d) Takeda 1,
(e) Takeda 2, (f) Takeda 3, (g) shear-slip 1, (h) shear-slip 2

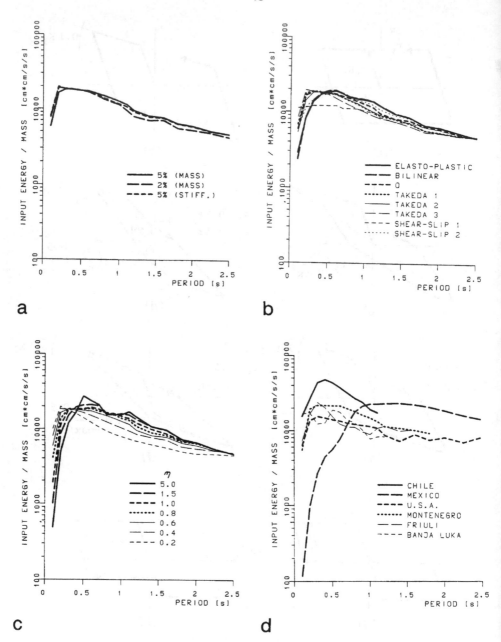

Figure 2. Mean input energy spectra (in the "average inelastic system" with 5 % mass-proportional damping, the parameters are varied one by one)
(a) influence of damping, (b) influence of hysteresis,
(c) influence of strength, (d) influence of input motion.
All input motions are scaled to $v_g \, t_D^{0.25} = 100$ where v_g is in cm/s and t_D is in seconds.

is defined as the work done by the equivalent force (mass multiplied by ground acceleration) on the equivalent fixed-base system. The effect of the rigid body translation of the structure is neglected.

The results of the parametric study have confirmed the conclusion drawn by some other investigators (e.g. Akiyama [10], Zahrah and Hall [14]) that maximum input energy per unit mass, which is imparted to systems with the natural period in the vicinity of the predominant period of ground motion, is a relatively stable parameter. It is (having in mind the large uncertainties associated with the majority of parameters in earthquake engineering) hardly dependent on damping (Fig. 2a), hysteresis (with the exception of the extreme shear-slip model, Fig. 2b) or the strength of the structure (with the exception of peak values in elastic structures, Fig. 2c). When comparing inelastic and elastic ($\eta = 5$) spectra, a smoothing effect of inelastic behaviour can be observed (Fig. 2c). A shift of the input energy curves to the shorter periods occurs as the strength and/or the "fatness" of the hysteresis loops decrease. This shift can be physically explained by a change in the effective period due to inelastic behaviour. It is larger in the case of strongly non-linear behaviour, typical for a system with lower strength and hysteresis with low "fatness".

The mean spectra of input energy per unit mass, based on constant ductility, are shown in Fig. 6a for the group representing "standard" ground motion (the U.S.A. and Montenegro groups). Three values of ductility ($\mu = 2$, 4 and 6) and two hysteretic models (bilinear and Q) were used. The input energy per unit mass for elastic system ($\mu = 1$) is shown for comparison. 5 per cent instantaneous-stiffness-proportional damping was assumed. It should be noted that in Fig. 6a a linear scale has been used on the energy axis, whereas in Fig. 2 a logarithmic scale was used. For this reason the peaks corresponding to the elastic system and to the system with low ductility (high strength) are more pronounced.

In contrast to the minor influence of structural parameters, shown in Figs. 2a, 2b, 2c and 6a, maximum input energy per unit mass is highly dependent on ground motion. Proper scaling of the accelerations should be used to eliminate this influence as far as possible. The scaling factor can be derived from formulae for the determination of input energy as a function of ground motion parameters. Such formulae have been proposed by Kuwamura and Galambos [15], Fajfar et al [2], Uang and Bertero [13] and Guo and Nishioka [16]. Akiyama [10] has proposed a simple procedure for determination input energy spectra based on the input energy imparted to elastic systems with 10 % damping. As an alternative, he has given simple formulae for input energy as a function of the intensity of earthquake and of the ground soil conditions. A detailed discussion of the different proposals is presented in paper [7]. Here only expressions from [15] and [2] will be shown.

After some transformations, the formula proposed by Kuwamura and Galambos [15] can be written in the form

$$\frac{E_I}{m} = \frac{1}{8} T_1 \int_0^{t_o} a(t)^2 \, dt \qquad (3)$$

where E_I is the input energy at the end of the excitation, m is the mass of the system, T_1 is the predominant period of the ground motion, a(t) is the ground acceleration as a function of the time t, and t_0 is the duration of the complete ground motion.

In Fig. 3a the relation (3) is compared with maximum values of input energy spectra for the "average inelastic system" (with mass-proportional damping) determined by dynamic analysis for the 72 different input motions listed in Table 1. T_1 in Eq. (3) was determined according to Eq. (2). It can be seen that generally Eq. (3) yields too low an estimate of the maximum input energy. A modified formula for input energy per unit mass is proposed

$$\frac{E_I}{m} = 0.85 \; \frac{v_g}{a_g} \; \int_0^{t_o} a(t)^2 \, dt \tag{3a}$$

In Eq. 3a it has been assumed that the predominant period T_1 is proportional to the peak ground velocity to acceleration ratio (see Eq. 2). The coefficient of variation is 0.4. It should be noted that the choice of another structural system would not influence the results noticeably.

According to [2], the maximum input energy per unit mass can be estimated from the formula

$$\frac{E_I}{m} = 2.2 \; t_D^{0.5} \, v_g^2 \tag{4}$$

where t_D is the duration of strong ground motion in seconds, according to Trifunac and Brady [18]. This empirical formula is the result of a statistical study, where 5 groups of ground motions were used (U.S.A., Montenegro, Friuli, Mexico and Banja Luka) (Fig. 2d). The mean coefficient of variation is about 0.5. Input energy per unit mass, computed according to Eq. (4), is plotted in Fig. 3b and compared with the maximum spectral values of the "average inelastic system". It can be seen that the dispersion of results is larger than in the case of Eq. (3a). Eq. (4) is apparently not suitable for the types of ground motion recorded in Chile, in Japan (group 2) and in Bucharest, i.e. for long duration ground motion with a short predominant period and for ground motion with a single large peak connected with a long pulse. In the first case the approximate results are too small and in the second case they are too large. In the case of usual ground motion, however, Eq. (4) can be easily used in design. Peak ground velocity is perhaps the most stable ground motion parameter, and the order of magnitude of the duration of ground motion can usually be easily predicted. This is not the case with the integral in Eq. (3).

The input energy according to Eqs. (3) and (4) represents the mean values of the maximum input energy which corresponds to a structure with a natural period in the vicinity of the predominant period of the ground motion T_1. Usually it is conservatively assumed ([10,15]) that input energy is constant in the whole medium- and long-period ranges (i.e. for all periods longer than T_1). The authors [4] assumed less conservatively that the input energy in the long-period region decreases with an increase of the natural period T.

In the short-period region input energy is highly dependent on various structural parameters. According to Akiyama [10] a linear relationship between equivalent velocity and

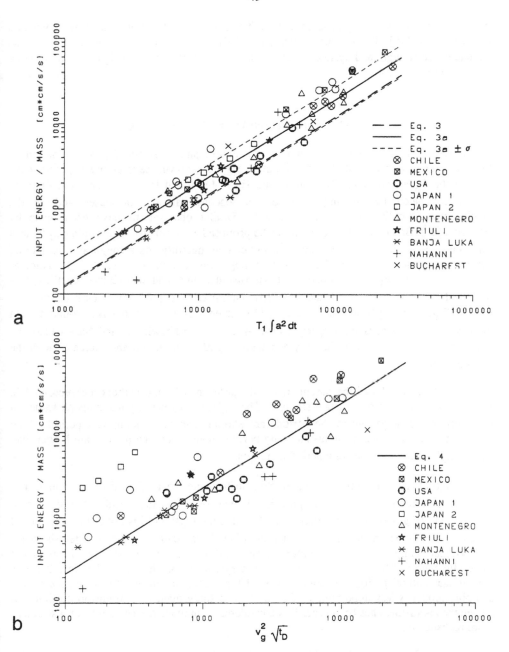

Figure 3. Maximum values of input energy per unit mass imparted to the
"average inelastic system" with 5 % instantaneous-stiffness-proportional
damping for the records listed in Table 1
(a) Comparison with Eqs. 3 and 3a   (b) Comparison with Eq. 4.

period can be used as a rough approximation. The same assumption was made by Kuwamura and Galambos [15]. The authors of this paper believe that a more realistic approximation dependent on strength and hysteretic behaviour will be needed for a design procedure based on the energy concept.

HYSTERETIC TO INPUT ENERGY RATIO

The hysteretic to input energy ratio is a convenient parameter for the determination of hysteretic energy, provided that the input energy is known. Several investigators have studied the E_H/E_I ratio recently. Zahrah and Hall [14] found that the proportion of input energy dissipated by yielding increases as viscous damping decreases and as the displacement ductility of a system increases. Similar observations have been made by Akiyama [10]. Based on these observations he proposed a formula for determining E_H/E_I as a function of viscous damping and strength rather than ductility. As a rough approximation, Akiyama proposed also a simplified equation, neglecting the influence of strength. Another expression for E_H/E_I as a function of viscous damping and cumulative ductility was proposed by Kuwamura and Galambos [15]. Nariyuki et al [18] have observed that the relation E_H/E_I versus period is not influenced by differences in the earthquake ground motion provided that the periods are properly scaled. According to Krawinkler and Nassar [19], the E_H/E_I ratio based on constant ductility decreases slightly with period and increases with the ductility ratio.

Selected results of the parametric study performed by the authors are presented in Fig. 4, which is based on constant strength. The E_H/E_I ratio generally has its peak values in the short-period range, where the periods are shorter than the predominant period of the ground motion T_1. In the medium- and long-period range, where the periods are longer than T_1, the E_H/E_I ratio decreases as a period increases.

One of the most important parameters influencing the E_H/E_I ratio is damping. The result of a decrease in viscous damping is an increase of the E_H/E_I ratio over the whole period range (Fig. 4a). It is important to realize that the E_H/E_I ratio is noticeably influenced by the mathematical modeling of viscous damping. In the case of inelastic analysis mass-proportional damping dissipates more energy and is thus more effective in reducing hysteretic energy. This tendency is larger for a system in the short-period range. Systems with longer natural periods experience less stiffness degradation and the influence of the damping model is less important. In the parametric study mainly damping proportional to instantaneous stiffness has been used. It is considered to simulate the behaviour of a structure more realistically.

The influence of different kinds of hysteretic behaviour can be observed in Fig. 4b. It can be seen that the type of hysteretic behaviour has a surprisingly small influence in the case of instantaneous-stiffness-proportional damping. This is not the case if mass-proportional damping is used [5]. An important difference can be observed only in the case of Takeda's hysteretic rules where trilinear envelopes were used. The period T in the figures corresponds to the initial stiffness, i.e. the stiffness before cracking in the case of a trilinear envelope. If an equivalent stiffness or the yield stiffness was considered, the difference between the two sets of curves, corresponding to bilinear and trilinear envelopes, would disappear.

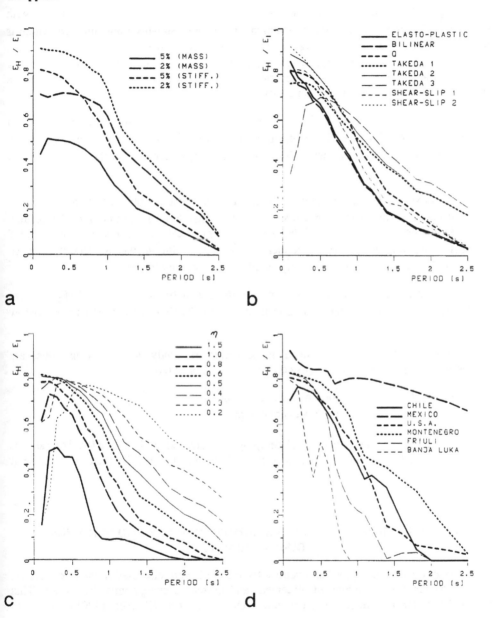

Figure 4. The mean ratio of hysteretic to input energy versus period
(in the "average inelastic system" with 5 % instantaneous-stiffness-proportional
damping, the parameters are varied one by one)
(a) influence of damping, (b) influence of hysteresis,
(c) influence of strength, (d) influence of input motion

The influence of the strength of a system can be seen in Fig. 4c. It is usually believed that E_H/E_I decreases with an increase in strength (or with a decrease in ductility). It was surprising to find that this is not always the case. The maximum E_H/E_I ratio is, with exception of a system with very high strength ($\eta > 1$), a reasonably stable quantity which is practically independent of the strength of the system. The decrease of the E_H/E_I ratio in the medium- and long-period range, however, depends strongly on strength. A system with low strength experiences large inelastic excursions and dissipates more hysteretic energy than a system with high strength and the same stiffness.

The mean values of E_H/E_I for different groups of records are shown in Fig. 4d. It can be seen that the maximum E_H/E_I values are more or less the same for all groups of records. The value of the predominant period T_1, however, strongly depends on the type of ground motion.

E_H/E_I versus T relations (mean values) based on the assumption of constant ductility were calculated for the group of records representing "standard" ground motion. The results, shown in Fig. 6c, indicate that the E_H/E_I ratio will be practically independent of the period T if constant ductility is assumed. An increase in E_H/E_I with increasing ductility can be observed. This increase, however, is very small, unless ductility is very low.

The coefficients of variation obtained in the parametric study of the E_H/E_I ratio are much smaller than those corresponding to input energy. Their value is in most cases below 0.1.

Based on observations obtained in the parametric study, the following values are proposed as an upper bound for the hysteretic to input energy ratio

$E_H/E_I = 0.8$ for 5 per cent damping
$E_H/E_I = 0.9$ for 2 per cent damping

In the case of mass-proportional damping the values should be smaller. The same is true if the strength parameter is large (e.g. $\eta \leq 1.0$) and/or the ductility factor is small (e.g. $\eta \leq 2$).

THE RELATION BETWEEN HYSTERETIC ENERGY AND MAXIMUM DISPLACEMENT

For the determination of equivalent ductility factors, which are discussed in the next chapter, the relation between maximum displacement and hysteretic energy must be known. This relation has so far not attracted much attention. Only a paper by Hirao et al [20] is known to the authors.

The first author [6] has introduced the dimensionless parameter γ

$$\gamma = \frac{\sqrt{\dfrac{E_H}{m}}}{\omega D} \tag{5}$$

where E_H is the dissipated hysteretic energy, D is the maximum displacement, m is the mass of the system and ω is its natural frequency. The parameter γ controls the reduction of the displacement ductility due to low-cycle fatigue, as shown in the next chapter.

Selected results of the parametric study on the parameter γ are shown in Figs. 5 and 6f.

The influence of damping is shown in Fig. 5a. It can be seen that the parameter γ is practically independent of both the mathematical modeling of the damping and of the damping coefficient.

As far as the influence of different hysteresis rules is concerned (Fig. 5b), bilinear and ideal shear-slip (model 1) hystereses represent the bounds of all investigated models (Fig. b). Hystereses simulating the flexural behaviour of steel structures (bilinear and elastoplastic) result in higher values of γ than those typical for the flexural behaviour of reinforced concrete structures (Q- and Takeda's models). It can be concluded that, generally, the value of γ increases with an increase in the "fatness" of the hysteretic loops.

The relatively large influence of the strength of a system can be seen in Fig. 5c. For code designed structures, high strength (typically $\eta \cong 1$) is required in the short-period region, whereas low strength (typically $\eta < 0.2$) meets the code requirements in the long-period region (see Fig. 6d). Having this fact in mind, it can be concluded from Fig. 5c that the γ - T relation of code designed structures can be reasonably approximated by the envelope of γ - T curves, which generally decreases with increasing period (with an exception in the case of bilinear systems in the short-period range). This observation, as well as a similar conclusion which can be drawn from other figures (e.g. Fig. 6f), suggests that low-cycle fatigue is a less important issue in the long-period range.

The influence of different ground motions can be observed in Fig. 5d. The major parameter which influences the value of the parameter γ seems to be the number of large-amplitude inelastic cycles a structure experiences during an earthquake. This number depends mainly on the duration of the strong ground motion, but other parameters, like the predominant period of the ground motion and the shape of response spectrum, may also have significant influence. The shift of maximum γ values towards longer periods, in the case of the Mexico City group, is a consequence of longer predominant periods.

The mean γ - T relations for bilinear and Q-hysteresis, for constant ductilities of $\mu = 2$, $\mu = 4$, and $\mu = 6$, for the group of records representing "standard" ground motion are shown in Fig. 6f. It can be seen again that bilinear hysteresis yields higher γ values than the Q-model in the whole period range except of very short periods. The parameter γ varies within relatively narrow limits, with a clear tendency to decrease with increasing periods. In the case of a bilinear hysteretic model, a decrease of γ can be observed in the short-period

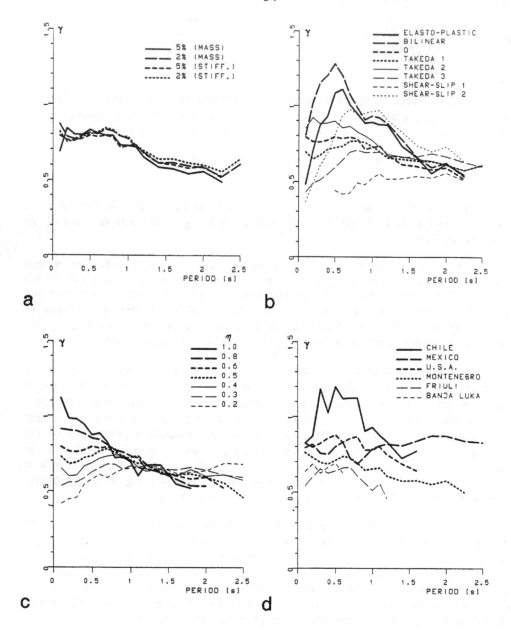

Figure 5. The mean values of the parameter γ versus period
(in the "average inelastic system" with 5 % instantaneous-stiffness-proportional damping,
the parameters are varied one by one)
(a) influence of damping, (b) influence of hysteresis,
(c) influence of strength, (d) influence of input motion

region, too. The parameter γ increases slightly with decreasing ductility. In the case of very low ductilities, however, an opposite trend can be observed. The value γ = 0 corresponds to elastic behaviour.

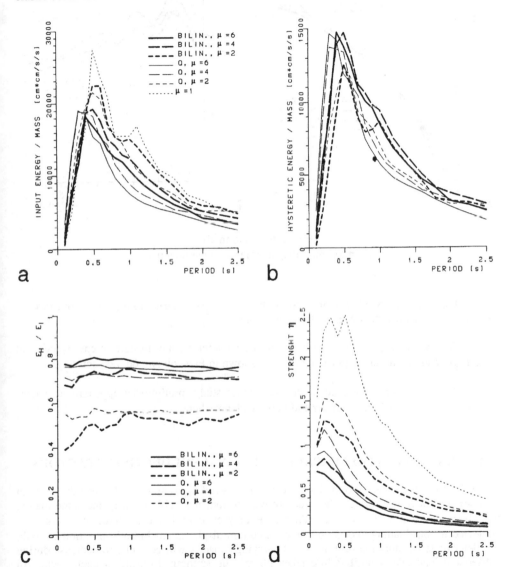

Figure 6. Mean values of response parameters, based on constant ductilities for the bilinear and Q-model, as functions of period. (The U.S.A. and Montenegro records represent the input motion. All records are scaled to $v_g \, t_D^{0.25} = 100$, where v_g is in cm/s and t_D is in seconds. 5 per cent instantaneous-stiffness-proportional damping is used.)
(a) input energy per unit mass, (b) hysteretic energy per unit mass
(c) hysteretic to input energy ratio, (d) strength

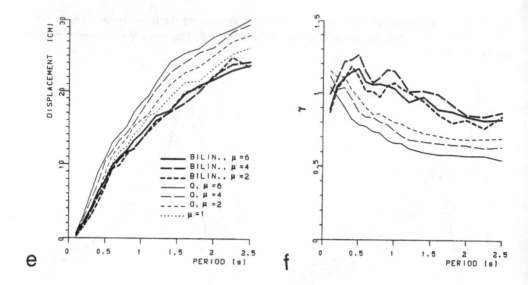

Figure 6. (continued)
(e) maximum displacement, (f) parameter γ

The coefficients of variation, computed in the parametric study, vary from case to case. They are, with very few exceptions, well below 0.3.

The two main parameters influencing the parameter γ, i.e. hysteretic energy per unit mass E_H/m and maximum displacement D, are shown in Figs. 6b and 6e, respectively.

Based on results obtained in the parametric study, preliminary approximate values will be proposed for the parameter γ and published elsewhere.

ENERGY DISSIPATION CAPACITY - EQUIVALENT DUCTILITY FACTORS

Under load reversals well into the inelastic range (beyond a certain critical level) the strength of a structure will deteriorate. The structure will no longer be able to carry the same load at the given deformation level. (According to Paulay et al [21], the strength of a structure, subjected to four loading cycles up to the maximum displacement in both directions, should not deteriorate by more than 20 %.) Load carrying capacity will continue to decrease during subsequent load cycles until failure takes place. The strength drop-off in each cycle will depend on the amount by which the critical strain has been exceeded. This phenomenon is generally known as low-cycle fatigue.

It follows from the above discussion that a substantial reduction in strength can be prevented by limiting the amplitude of cyclic deformations. The allowable maximum displacements and ductility (supply) depend on the number of cycles. Many cycles of small displacement amplitude will produce damage similar to that resulting from a large amplitude of monotonic deformation. It can be concluded that the deformation capacity of a structure is reduced as a consequence of the dissipation of hysteretic energy caused by cyclic load reversals. Consequently, a reduced ductility supply can be defined, which reflects the influence of cyclic response, and which should be used instead of the conventional monotonic ductility supply in design procedures. The concept of equivalent ductility is based on the assumption that the damage produced by a number of cycles at an equivalent (reduced) amplitude will be equal to the damage due to monotonic deformation of a large amplitude.

The idea of equivalent ductility factors has been already used, in one form or another, by several researchers [22,24]. In [6] it was elaborated in a form appropriate for direct practical application. Three different failure hypotheses were used and simple formulae for the calculation of equivalent ductility factors were derived.

In design procedures only collapse limit states are usually considered. A more rational design procedure should permit the designer to choose explicitly a damage limit state that is compatible with the return period of the design earthquake motion [24]. This idea has been realized by including damage indices explicitly into the formulae for determining equivalent ductility factors.

Equivalent ductility factors based on three damage models
The structure is simulated by an equivalent SDOF ideal elastoplastic system. The ductility factor μ is defined as the ratio of the actual displacement D to the yield displacement D_y $(\mu = D / D_y)$. It is assumed that the actual displacement is equal to the displacement which corresponds to the permissible damage index. Consequently, the actual ductility factor is equal to the equivalent ductility factor.

In the following text expressions for equivalent ductility factors will be given for three different failure hypotheses. Two of them define upper and lower bounds for the possible range of ductility factors. The third one is considered to produce more realistic values, than those predicted by the first two hypotheses.

A) The usual assumption, which does not take into account the low-cycle effect, is that failure is due to the maximum plastic displacement. If displacement supply is based on the ultimate displacement under monotonic loading, this assumption will not be conservative and ductility factors based on it will represent the upper limit of the possible range of factors. The damage index DM can be written in the form [25,26]

$$DM = \frac{D - D_y}{D_u - D_y} = \frac{\mu - 1}{\mu_u - 1} \tag{6}$$

where D_u is the ultimate displacement under monotonic loading and μ_u is the corresponding ductility factor. A damage index which is equal to or larger than unity implies failure. From Eq. (8) the following relation between the equivalent ductility factor μ and the ultimate monotonic ductility factor μ_u can be obtained

$$\mu = DM (\mu_u - 1) + 1 \tag{7}$$

B)   According to a very conservative assumption, the capacity of a structural system to dissipate hysteretic energy under cyclic loading is equal to its energy dissipation capacity under monotonic loading. The value of the equivalent ductility factor based on this assumption may be considered to be the lower bound value. In an ideal elastoplastic system (or in an equivalent elastoplastic system with the same energy dissipation capacity as an actual system) the damage index may be written in the form

$$DM = \frac{E_H}{F_y(D_u - D_y)} = \frac{E_H}{F_y D_y(\mu_u - 1)} \tag{8}$$

After some transformations (see [5]) a relation between the equivalent ductility factor μ and the ultimate monotonic ductility factor μ_u can be obtained

$$\mu = \frac{\sqrt{DM(\mu_u - 1)}}{\gamma} \tag{9}$$

where γ is defined according to Eq. (5)

Eq. (9) suggests that the reduction of ductility due to low-cycle-fatigue is controlled by the dimensionless parameter γ which was discussed in the previous chapter. A large reduction of monotonic ductility, representing an important effect of low-cycle fatigue, will be obtained if γ is large. If γ is small, however, the low-cycle fatigue effect will become negligible, and the usual displacement ductility will control damage. The parameter γ will attain its minimum value γ_{min} if the system experiences only one inelastic half-cycle. For the ideal elastoplastic system the following expression can be derived

$$\gamma_{min} = \frac{\sqrt{\mu - 1}}{\mu} \tag{10}$$

If the value for the parameter γ according to Eq. (10) is substituted into Eq. (9) the expected result $\mu = \mu_u$ will be obtained in the case DM = 1. In the case of elastic behaviour hysteretic energy vanishes and γ becomes zero.

C)   A damage model which is more realistic than the models defined in Eqs. (6) and (8), and which is suitable for the derivation of an equivalent ductility factor, is the Park-Ang model [27]

$$DM = \frac{D}{D_u} + \beta \frac{E_H}{F_y D_u} = \frac{\mu}{\mu_u} + \beta \frac{E_H}{F_y D_y \mu_u} \tag{11}$$

where β is a constant which depends on the structural characteristics. The parameter β may be interpreted as a parameter controlling strength deterioration as a function of the amount of dissipated energy (Park et al [28]). Cosenza et al [26] have found that the value $\beta = 0.15$ correlates closely to results based on the damage models proposed by Banon et al [29] and by Krawinkler [30]. Consequently, $\beta = 0.15$ seems to be a realistic mean value. The relation between β for structural elements and β for the entire structure is still to be investigated.

The Park-Ang model is a linear combination of simplified versions of the models defined in Eqs. (6) and (8). In spite of some deficiencies indicated by Chung et al [31], it has been widely used in recent years. Its main advantages are simplicity and a broad experimental basis. It includes hysteretic energy (rather than the number of cycles), which is an advantage for the concept used in this paper.

From (11) and (5) the following relations between the equivalent ductility factor and the ultimate monotonic ductility factor can be obtained

$$\frac{\mu}{\mu_u} = \frac{DM}{1 + \beta \gamma^2 \mu} \tag{12}$$

or

$$\mu = \frac{\sqrt{1 + 4\,DM\,\beta\,\gamma^2\,\mu_u} - 1}{2\,\beta\,\gamma^2} \tag{13}$$

Eq. (12) suggests that the reduction of ductility due to low-cycle fatigue is controlled by the parameters β and γ, as well as by the actual amplitude of vibration (expressed in terms of μ). The permissible damage index DM, which will be smaller than 1 in the case of a ground motion with a short return period, is explicitly included in formula.

CONCLUSIONS

Seismic energy demand, in terms of input energy per unit mass, the hysteretic to input energy ratio, and the parameter γ have been studied. The following main conclusions and results have been obtained.

1. Maximum input energy is imparted to systems with natural periods in the vicinity of the predominant period of the ground motion T_1. This is a relatively stable quantity (considering the inherent uncertainty of the majority of parameters in earthquake engineering), reasonably independent of structural parameters and of ground motion provided that proper scaling is used.

2. A reasonable estimate of maximum input energy can be obtained from the formula proposed by Kuwamura and Galambos and modified by the authors (Eq. 3a). A simple empirical formula has been proposed by the authors (Eq. 4). It is based on two standard ground motion parameters, and can be easily used in design procedures if the usual types if ground motions are expected.

3. The input energy imparted to systems with natural periods in the short-period range ($T < T_1$) depends strongly on different structural parameters. For a system with a fundamental natural period in the long-period range, which generally experiences only a small number of large amplitude cycles, energy demand is believed to be less important. Maximum displacement governs the design of such structures. It should be noted that the

periods defining the short- and the long-period range depend on the frequency content of the ground motion.

4. E_H/E_I is the most stable of the investigated parameters. It is not much influenced by different structural and ground motion parameters with the exception of damping. The coefficients of variation determined in statistical studies are very low (below 0.1 in most cases). Approximate upper bound values for this parameter have been proposed.

5. The dimensionless parameter γ is a convenient parameter for expressing the relation between dissipated energy and maximum displacement. It controls the reduction of deformation capacity due to dissipated hysteretic energy. The parameter γ depends mainly on the type of ground motion and on hysteretic behaviour, as well as on the natural period of a structure. A moderate influence of ductility and/or strength can also be observed. Hopefully it will be possible to propose reliable approximate values for the parameter γ, as a function of the basic structural and ground motion parameters to be used in design.

An approach including cumulative damage indicators, e.g. hysteretic energy, is not without difficulties in terms of its practical application, especially on the supply side. A promising technique for handling the problem seems to be the introduction of an equivalent ductility factor which takes into consideration the influence of cyclic load reversals. Such an approach represents a minor adjustment to the well-known concept of ductility factors that is relatively well understood and widely employed in practice. Reduction (or behaviour, R, q) factors taking into account low-cycle fatigue and a prescribed damage state can be easily determined by using equivalent ductility factors instead of conventional ductility factors.

A realistic value of the equivalent ductility factor lies between two extreme values representing two extreme failure hypotheses: (a) failure is due to the maximum plastic displacement; displacement supply is equal to the ultimate displacement under monotonic loading; (b) failure is due to dissipated energy; the capacity of a structural system to dissipate hysteretic energy under cyclic loading is equal to its energy dissipation capacity under monotonic loading. A failure hypothesis, which yields equivalent ductility factors between two extreme cases, is based on the Park-Ang damage model. Equivalent ductility factors, derived by using this hypothesis, depend inter alia on the parameter β, which is connected with the energy dissipation capacity of a structure. A lot of research is still needed to obtain reliable β-values for different structural elements and systems.

ACKNOWLEDGEMENTS

The results presented in this paper are based on work supported by the Ministry for Science and Technology of the Republic of Slovenia and by the U.S. - Yugoslav Joint Fund for Scientific and Technological Cooperation.

REFERENCES

1. Newmark, N.M. and Hall, W.J., Earthquake spectra and design. Earthquake Engineering Research Institute, Berkeley, CA, 1982.

2. Fajfar, P., Vidic, T. and Fischinger, M., Seismic demand in medium- and long-period structures, Earthquake engng. and struct. dyn., 1989, **18**, 1133-1144.

3. Fajfar, P., Vidic, T. and Fischinger, M., A measure of earthquake motion capacity to damage medium-period structures, Soil dyn. earthquake eng., 1990, **9**, 236-242.

4. Fajfar, P. and Fischinger, M., A seismic design procedure including energy concept, Proc. 9th European conf. earthquake eng., Moscow, 1990, Vol. 2, pp. 312-321.

5. Vidic, T., Fajfar, P. and Fischinger, M., On the hysteretic to input energy ratio, Proc. 6th Canadian conf. earthquake eng., Toronto, 1991, pp. 69-76.

6. Fajfar, P. Equivalent ductility factors taking into account low-cycle fatigue, Submitted for publication.

7. Fajfar, P., Vidic, T. and Fischinger, M., On the energy input into structures, Pacific conf. earthquake eng., Auckland, New Zealand, 1991, accepted for publication.

8. Saiidi, M. and Sozen, M.A., Simple nonlinear seismic analysis of RC structures, J. struct. div., ASCE, 1981, **107**, 937-952.

9. Fajfar, P. and Fischinger, M., N2 - a method for non-linear seismic analysis of regular buildings, Proc. 9th world conf. earthquake eng., Tokyo, Kyoto, 1988, Vol. 5, pp. 111-116.

10. Akiyama, H., Earthquake-resistant limit-state design for buildings. University of Tokyo Press, 1985.

11. Heidebrecht, A.C., Private communication, 1987.

12. Otani, S., Hysteresis models of reinforced concrete for earthquake response analysis, Journal of the Faculty of Engineering. The University of Tokyo, 1981, **36**, pp. 125-159.

13. Uang, C.M. and Bertero, V.V., Evaluation of seismic energy in structures. Earthquake enging. and struct. dyn., 1990, **19**, 77-90.

14. Zahrah, T.F. and Hall, W.J., Seismic energy absorption in simple structures. Structural Research Series No. 501, Civil Engineering Studies, University of Illinois, Urbana-Champain, Illinois ,1982.

15. Kuwamura, H. and Galambos, T.V., Earthquake load for structural reliability. J. of struct. eng., ASCE, 1989, **115**, 1446-1462.

16. Guo, X.Q. and Nishioka, T., Estimation of total input energy to structures under earthquakes, Proc. of the Japan Society of Civ. Eng., 1989, No 410/I-12, 405-413, in Japanese with english summary.

17. Trifunac, M.D. and Brady, A.G., A study of the duration of strong earthquake ground motion. Bull of the seism. soc. of Am., 1975, **65**, 581-626.

18. Nariyuki, Y., Hirao, K. and Ohgishi, K., Study on relation between Fourier spectra of earthquake motion & energy response spectra of SDOF systems. Proc. 2nd second East Asia-Pacific Conference on Structural Engineering & Constructions, Chiang Mai, 1989, pp. 1503-1509.

19. Krawinkler, H. and Nassar, A., Damage potential of earthquake ground motions, Proc. 4th U.S. conf. earthquake eng., Palm Springs, California, EERI, 1990, Vol. 2, 945-954.

20. Hirao, K., Nariyuki, Y., Sawada, T. and Sasada, S., On the relation between maximum displacement and hysteretic energy of SDOF structures under strong earthquake motion, Proc. 2nd East Asia-Pacific conf. struct. eng. & construction, Chiang Mai, 1989, pp. 1628-1633.

21. Paulay, T., Bachmann H. and Moser, K., Erdbebenbemessungen von Stahl-betonhochbauten, Birkhauser Verlag, Basel, 1990.

22. Lashari, B., Seismic risk evaluation of steel structures based on low-cycle fatigue. Reliability Engng. and System Safety, 1988, **20**, 297-302.

23. McCabe, S.L. and Hall, W.J., Assessment of Seismic Structural damage, J. struct. engng., ASCE, 1989, **115**, 2166-2183.

24. Bracci, J.M., Reinhorn, A.M., Mander, J.B. and Kunnath, S.K., Deterministic model for seismic damage evaluation of reinforced concrete structures, Technical Report NCEER-89-0033, National Center for Earthquake Engineering Research, State University of New York at Buffalo, 1989.

25. Powell, G. H. and Allahabadi, R., Seismic damage prediction by deterministic methods: concepts and procedures, Earthquake engng. and struct. dyn., 1988, **16**, 719-734.

26. Cosenza, E., Manfredi, G. and Ramasco, K., An evaluation of the use of damage functionals in earthquake-resistant design, Proc. 9th European conf. earthqake eng., Moscow, 1990, Vol. 9, pp. 303-312.

27. Park, Y. J., Ang, A. H.-S. and Wen, Y. K., Seismic damage analysis and damage-limiting design of R.C. buildings, Structural Research Series No. 516, University of Illinois, Urbana, Illinois, 1984.

28. Park, T. Y., Reinhorn, A. M. and Kunnath, S. K., IDARC: Inelastic damage analysis of reinforced concrete frame-shear-wall structures, Technical Report NCEER-87-0008, National Center for Earthquake Engineering Research, State University of New York at Buffalo, 1987.

29. Banon, H., Biggs, J. and Irvine, H., Seismic damage in reinforced concrete frames, J. struct. div., ASCE, 1981, **107**, 1713-1729.

30. Krawinkler, H., Performance assessment of steel components, Earthquake spectra, 1987, **3**, 27-41.

31. Chung, Y. S., Meyer, C. and Shinozuka, M., Seismic damage assessment of reinforced concrete members, Technical Report NCEER-87-0022, Buffalo, NY, 1987.

SEISMIC DESIGN OF STRUCTURES FOR DAMAGE CONTROL

ANDREI M. REINHORN   SASHI K. KUNNATH   JOHN B. MANDER
State University of New York at Buffalo
Buffalo, New York, 14260 USA

ABSTRACT

The objective of current seismic design practice for structures is related to providing life safety during severe ground motion primarily, providing safety and functionality of contents secondly and only in the end provide structure survivability. These priorities are often changed, depending on the functional design objectives, investments, maintenance costs and repair capabilities. In the structural design some degrees of damage are allowed to occur depending on the seismic risks and the expected earthquake magnitudes. However, the life safety of inhabitants, the integrity and functionality of buildings' contents are dependent on the performance and the survivability of entire structural system. Therefore the degree of acceptable damage in the structural systems and its control during and after a seismic event are of foremost importance.

The current paper summarizes the state of knowledge related to the definitions of damage states and quantifiable damage indicators, useful for characterizations of degrees of damage of structural members. Links between damage states and damage indices are established and used to develop suitable strength reduction factors which target the control of degrees of damage. Although the state-of-the-art in qualification of degrees of seismic damage is in its beginning stages, the principles developed herein to determine design procedures to control this damage may be valid for future developments.

INTRODUCTION

The current approach in the design of earthquake resistant structures is based on damage prevention during low magnitude earthquakes, allowing some damage during moderate or intermediate tremors and on prevention of collapse during severe earthquakes. This concept seem more economical and more feasible than complete damage prevention in all cases. The concept is implemented by allowing for a design in which the structural members will experience post-elastic excursions associated with permanent deformations, stiffness and strength deteriorations. A member experiencing such transformations can be defined as damaged.

Indeed after a strong earthquake, many buildings experience various degrees of damage and some collapse. One of the most difficult tasks of the post earthquake inspection is to asses and quantify the seismic damage or estimate the seismic safety and further usability of the remaining building stock. The classification of damage, empirical at its base, is different from one country to another or from an engineering school to another. A common factor however unifies these schools. The need for a reliable way to mitigate the hazards to life and property during and in the aftermath of an earthquake. Therefore a better understanding of various degrees of damage and control of such damage by proper design of structures becomes an essential task to the engineering community.

Current codes of design require to determine the strength of members such that would resist a fraction of the strength required for an expected pure elastic response. As such the designer relies on the capability of members to experience ductile behavior and dissipate energy without collapse. The strength reduction factors are therefore related to the capability of members to develop substantial ductilities and prevent damage.

Krawinkler et al. (1990), suggested correlations between the strength reduction factors and the ductility demands based on actual earthquake characteristics and an ultimate design criterion. Ang H.S. (1988) established relations for ductility requirements and the base shear coefficients using a limit design criterion based on one, single degree of damage. McCabe and Hall (1989) went one step further, to establish spectra for demanded ductility capacity and the code displacement design coefficients based on various degrees of damage. Current paper defines (i) degrees of damage in terms of further usability of structures, (ii) some response indicators correlated with such degrees of damage, and (iii) attempts to determine a correlation between strength reduction factors and the required ductility of members correlated to various degrees of damage.

DAMAGE STATES

The quantification of damage for the sake of classification it is a very difficult task. Very few recommendations are currently available. Anagnostopoulus, Petrovski and Bouwkamp (1989) suggest a damage and usability classification based on three categories primarily suitable to reinforced concrete or masonry structures (see Col. 1 in Table 1). The suggested categories are (i) usable structures representing undamaged or slightly damaged structures; (ii) temporarily unusable representing moderate to heavily damaged structures; (iii) unusable representing severely damaged to partial or totally collapsed structures. These categories are defined and based on visibly observed cracking, without or with structural meaning, deterioration of concrete members and masonry walls, spalling of concrete cover and buckling of steel reinforcement, damage of important and significant members of the structural system. While the damage is only qualitatively estimated, the damage states and the usability categories can be used for classifications and further decision making.

Park, Ang and Wen (1985) suggest five degrees of damage related to their physical appearance (see Col. 2 in Table 1). They related the degrees of damage to an overall damage indicator which can be used to define the usability of buildings by assigning a repairability (or irreparability) limit, ie. 0.4 on a scale of 1.0 in which the upper limit defines total collapse. The above damage indicator is making an attempt to quantify the damage by means of studying the deformations and the current strength of members in comparison to ultimate limits of such quantities.

Based on correlations of degrees of damage of various concrete members, their appearance and usability, a new scale of damage states related to their service status is defined (see column (3) in Table 1) for reinforced concrete beams and columns subjected to cyclic loading. The scale was also earlier calibrated for nine concrete buildings that were moderately or severely damaged during the 1971 San Fernando earthquake. (Park et al.(1985)). The *serviceability state* suggests that only minor repairs, usually cosmetic, are necessary. The *repairability state* suggests that serious repairs are required to restore the structure to a suitable state. The *irreparability state* implies that repair cannot restore structure's functionality, however, the structure did not collapsed and possibly life could have been saved.

Table 1 - Correlation of Damage Indices and Damage States

USABILITY[1]	DEGREE OF DAMAGE[2]	DAMAGE (SERVICE) STATE	LIMIT STATE DAMAGE INDEX	APPEARANCE[3]
(1)	(2)	(3)	(4)	(5)
		Undamaged		Undeformed/uncracked
			0.00	
Usable	Slight	Serviceable		Moderate to severe cracking
			0.20-0.30	
	Minor	Repairable		Spalling of concrete cover
Temporarily-	Moderate		0.50-0.60	
Unusable	Severe	Irreparable		Buckled bars, exposed core
			> 1.00	
Unusable	Collapse	Collapse		Loss of shear/axial capacity

[1] Acc. to Anagnostopoulus et al. (1989)
[2] Acc. to Park et al. (1985)
[3] Acc. to Bracci et al. (1989)

QUANTIFICATION OF DAMAGE

In the study of structural damage, a large number of indicators were proposed. Chung, Meyer an Shinozuka (1987) summarized the state-of-the-art, pointing to definitions of damage indicators for individual members (or local) and overall indicators (or global). Several new indicators were proposed after 1987 based on similar approaches. A brief review is presented in the following for the sake of completion.

Most of the indicators compare achieved performance of members to the largest capacity of same performance quantity. Bertero and Bressler (1977) suggest a weighted ratio of deformation response (demand) versus capacity of members as local damage indicators. This index relates almost entirely to the ductility of members without relation to the strength and stiffness deteriorations of beams and columns. Bertero et al. (1977) suggests the concept of weighted accumulations of damage in the evaluation of global indicators from the local ones, based on their importance and service history. Lybas and Sozen (1977), Roufaiel and Meyer (1985), DiPasquale and Cakmak (1989) and others, suggest local and global damage indices which compare actual stiffness degradation with maximum possible stiffness degradation. However depending on the definition of such stiffness (either equivalent or actual) the indicator has limited sensitivity in case of severe strength deterioration due to repetitive loading as occurs during earthquakes. As an alternative (or in addition), Blejwas and Bressler (1979), Krawinkler and Zohrei (1983), Roufaiel and Meyer (1985) and others monitor the actual permanent plastic deformations in comparison to the largest permanent deformations of members before collapse in order to quantify the damage.

Gosain (1977), Hwang and Scribner (1984), Darwin and Nmai (1986), Meskouris and Kratzig (1987), McCabe and Hall (1989) use a ratio of the work done or the energy consumed during a seismic event in respect to the total energy capacity of the member. These indicators produce good results incorporating the influences of both strength and deformations of members, however have some limitations in recognizing damage due to small inelastic excursions and influence of repetitive loading on the strength and stiffness deterioration of members. Banon, Biggs and Irvine (1981), Park Ang and Wen (1985), Bracci et al. (1989) use an index which consider both deformation ratios and energy ratios to quantify the members' damage. These indicators imply information on strength and stiffness deterioration and member fatigue. Chung, Meyer and Shinozuka (1987) suggest an index based on an accelerated Miner's rule based on low cycle fatigue in which the actual working cycles and the fatigue cycles are derived from a member hysteretic rule.

This index, although most complex, has a limited direct correlation to the observed parameters of a member during severe load reversals, such as permanent deformations and strength degradation.

A NORMALIZED DAMAGE INDICATOR

A newly proposed damage model utilizes the concepts of *consumption* and available *damage potential*. Damage potential (D_p) is the total capacity of the component to sustain damage. Damage consumption (D_c) is that portion of the available capacity that is lost or dissipated during the course of the applied load history. Failure of a component by purely monotonic loading represents an upper bound behavior. At the other extreme is the low cycle fatigue at a given amplitude of deformation, which represents a lower bound, as shown in Fig. 1.

The damage potential, D_p, of an RC structural component is defined as the total area enclosed by the monotonic, $f_m(\phi)$, and the failure, $f_f(\phi)$, envelopes:

$$D_p = \int_{-\phi_u}^{+\phi_u} \{f_m(\phi) - f_f(\phi)\} d\phi \tag{1}$$

where ϕ is the curvature and ϕ_u is the ultimate curvature.

Consider a RC component for which the results of a cyclic test are available. (See Fig.1a). A new curve defines the current damage level of the component, $f_c(\phi)$. This function represents a dynamic upper-bound envelope that is constantly dropping as a consequence of inelastic cyclic deformation. The positive and negative unloading stiffness paths delimit the unrecoverable deformation.

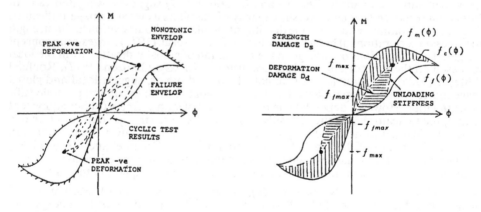

(a) Cyclic Test Data   (b) Damage Estimation

Fig.1. Conceptual Model of Damage

Two components of damage are defined as follows (Fig.1b): the first corresponding to strength-loss and the second arising from deformation related damage. Strength damage, D_s, is defined as the loss of damage potential due to strength deterioration and hysteretic dissipated energy. Deformation damage, D_d, accounts for the irrecoverable permanent deformations:

$$D_s = \int_{-\phi_u}^{+\phi_u} \{f_m(\phi) - f_c(\phi)\} d\phi, \text{ and } D_d = \int_{-\phi_a}^{+\phi_a} \{f_c(\phi) - f_f(\phi)\} d\phi \quad (2)$$

where ϕ_a represents the line joining f_{max} and f_{fmax} in Fig.1b. The cumulative effect of strength damage and deformation damage is termed *damage consumption*. A structural Damage Index (DI) is, therefore, defined as the ratio of the damage consumption to the damage potential:

$$DI = (D_s + D_d)/D_p \quad (3)$$

Application of Damage Model to RC Members. Consider a component whose force-deformation history at member ends is known. The peak deformation attained is $M_{\phi max}$, $\phi_{\phi max}$, as shown in Fig.2. The force and deformation at yield and ultimate levels for monotonic loading are established either through the use of empirical equations or some micro-modeling scheme.

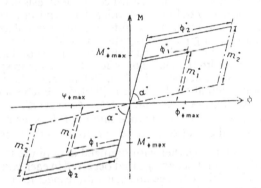

Fig.2. Implementation of Model for Bi-linear Idealization

The inelastic failure envelope should come directly form experimental testing. However, experimental data is limited. Consequently, only theoretical possibilities based on observed patterns of fatigue failure for metals, and some limited data on concrete, can be postulated (Bracci et al., 1989). One possibility is to assume the backbone (upper bound) curve for a RC member can be idealized by a bi-linear relationship, with a transposed form of the monotonic yield surface forming the lower bound curve.

Using this bi-linear idealization with the same envelope characteristics in compression and tension, the following simplified expression results for the Damage Index:

$$DI = D_2 + D_1(1 - D_2) \quad (4)$$

where: $D_1 = \phi_1/\phi_2$ and $D_2 = (m_2 - m_1)/m_2$ with m_1, m_2, ϕ_1, ϕ_2 as shown in Fig.2. (In terms of ductilities, $D_1 = (\mu_D - 1)/(\mu_C - 1)$ where μ_C is the ductility capacity corresponding to $\mu_C = \phi_2/\phi_Y$ and μ_D is the ductility demand corresponding to $\mu_D = \phi_1/\phi_Y$). When carrying out a damage analysis of a member, D_2, relates to the strength loss ($\Delta M = m_2 - m_1$) due to cyclic loading may be evaluated using the following relationship:

$$\Delta M = k \int dE / \phi_y \tag{5}$$

where ϕ_y is the yield curvature (or displacement), $\int dE$ is the total energy absorbed by the member (equal to the sum of the area enclosed within the hysteresis curves) and k is a strength deterioration factor. From a regression analysis of reinforced concrete members with differing amounts of longitudinal (ρ_l) and transverse reinforcement ratios (ρ_w), and axial load (P_e), an empirical strength deterioration relationship is found in the form:

$$k = 0.00857 \left(1 + 12 \frac{P_e}{f_c' A_g} \right) \left(1 - 0.5 \frac{\rho_w f_{yw}}{0.85 f_c'} \right) \left(1 - \frac{\rho_l f_{yl}}{0.85 f_c'} \right) \tag{6}$$

A 24% coefficient of variation between experimental and empirical prediction was obtained.

Damage Model Verification Using Observed Test Results. In order to investigate the validity of the different failure modes predicted by the proposed damage model, it was considered necessary to verify the model against component tests in which the experimental failure point could be precisely defined. Mander, Park and Priestley (1983) tested near full-sized hollow column (piers) specimens under varying levels of axial load with differing amounts of confining steel. A good photographic record of the tests at various levels of ductility was also available, thus enabling a visual description of damage at various stages of loading to be correlated with the Damage Index. Each column specimen had the same height, cross section and wall thickness of 3.2 m, 750 mm and 120 mm, respectively, and a constant amount of longitudinal reinforcement (0.0155). Each specimen was tested in a quasi-static fashion consisting of two complete cycles each at displacement ductility factors of $\mu = \pm 2, \pm 4, \pm 6,$ and ± 8, unless premature failure of the specimen occurred. Table 2 presents the evaluation of the control parameters based on the specimen characteristics.

Using the control parameters together with digitized data of the force-displacement hysteresis loops (Fig. 3), the amount of damage that occurred in each component was analyzed. Fig. 3 presents the results of the damage analysis with respect to cumulative displacement ductility, $\mu = \sum \Delta_a / \Delta_y$ (where $\sum \Delta_a$ is the sum of the total displacements) for Columns A, C and D, respectively.

Table 2. Results of Damage Analysis

Column	Control parameters used in damage analysis					Results of damage analysis		
	P_e/f_cA_g	ρ_w	Δ_y	Δ_u	R	D_1	D_2	DI
		%	(mm)	(mm)	Eq (6)	Eq (4)	Eq (4)	Eq (4)
A	0.1	0.21	14	260	0.0067	0.5	1.00	1.0
C	0.3	0.31	13	150	0.0112	1.0	0.69	1.0
D	0.3	0.21	13	90	0.0123	1.0	0.20	1.0

Damage analysis results of three of the columns tested by Mander et al. (1983) are summarized in Table 2. The results clearly show that the members failed due to the combined effects of deformation (D_1) and strength loss damage (D_2) caused by cyclic

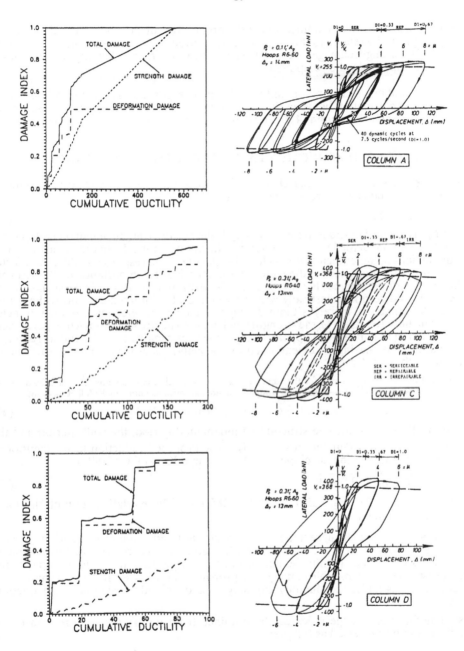

Fig.3. Damage Analysis of tested columns (Ref.[15])

loading, which confirms the foundation of the proposed damage model. The deformation damage is primarily related to the ultimate deformation and to some extent the unloading stiffness. The strength damage is related to strength loss from cyclic loading and the amount of energy dissipated by the component.

Based on the above study and on additional similar experimental results and numerical simulations, a correlation between the damage index and the damage states is suggested and shown in Table 1 (column (4)). These damage limits can be used as design targets.

STRENGTH REDUCTION FACTORS FOR DAMAGE CONTROL

Many of new seismic codes (NEHRP, (1989)) adopted the strength capacity requirements as a function of the elastic strength demand, ie:

$$F_D = F_{EL} / R_D \qquad (7)$$

where F_D is the design strength capacity, F_{EL} is the elastic strength demand and R_D is the strength reduction factor (or the response modification factor). It should be noted that in this discussions no other factors are included explicitly in the response modification factor such as overstrength, uncertainties, etc. To emphasize the importance of the strength reduction factor let assume that no reduction is permitted (ie. $R_D = 1$). In such case the structure would respond elastically without yielding of reinforcement and without permanent deformations, merely without damage. However, selecting a reduction factor larger than unity (ie. $R_D > 1$), the strength level of the structure will be such that permanent deformations and damage will appear. Only structures capable to withstand large plastic deformations (ie. having sufficient ductility capacity) may withstand the earthquake demands with little damage.

Hidalgo and Arias (1990) prepared for the Chilean seismic code a relation which links the ductility demands with the strength reduction factors based on study of earthquake response done by Ridell et al. (1989).

$$R_D = 1 + (T / T_o) / [0.1 + (T / T_o) / (\mu_D - 1)] \qquad (8)$$

where T, T_0 and μ_D are the structure's fundamental period, the soil's period and the displacement ductility demand, respectively. The proposed reduction factor mentioned above fits also the relation developed by Osteraas and Krawinkler (1990):

$$R_D = \sqrt[\alpha]{C (\mu_D - 1) + 1} \qquad (9)$$

where $C = 1.35 + .211 / T$ and $\alpha = .972 + .185 / T$. These relations fit both short and long period structures.

Osteraas and Krawinkler suggest the use of the strength reduction factors and the ratio of local to global overstrength such that the ductility capacity will exceed the ductility demand. It is obvious that if the ductility demand is equal or larger than the ductility capacity the structure will suffer irreparable damage and probably will collapse. If damage control and reduction is desirable, the ductility capacity should exceed substantially the ductility demand.

Assuming that the damage index can be obtained from Eq.(4), the term related to the ductilities can be expressed as follows:

$$D_1 = (D.I. - D_2) / (1 - D_2) \quad \ldots = \ldots \quad \frac{(\mu_D - 1)}{(\mu_c - 1)} = \frac{(D.I - D_2)}{(1 - D_2)} \quad (10)$$

Using the expressions in Eq (9) and in Eq (10), the strength reduction factor can be obtained as a function of the ductility capacity as follows:

$$R_D = \sqrt[\alpha]{C\,(\mu_c - 1)\,(D.I. - D_2)\,/\,(1 - D_2) + 1} \tag{11}$$

The relation above indicates that the strength reduction factor is a function of the damage indicator, D.I., and the strength deterioration (or energy absorption) factor, D_2.

In the design process the damage index, D.I., can be chosen to fit a desired targeted damage limit (as defined in Table 1) while the energy absorption factor, D_2, can be estimated for the structural system from the ground motion parameters. Therefore the ductility capacity, μ_c, and the strength reduction factor R_D become explicitly correlated.

Inspecting the relation in Eq (11) it is possible to note that the strength reduction factor R_D tends to unity for a performance without damage (D.I. = 0 and D_2 = 0). However, for a targeted collapse, D.I. = 1, for which the ductility capacity equals the ductility demand, $\mu_c = \mu_D$ independently of the energy dissipation, the strength reduction factor is reduced to the original relation in Eq.(9).

The relation in Eq. (9) is compared in Fig. 4 to the results of earthquake performances of typical bridge piers, having various ductility capacities and subjected to simulated ground motion. The simulations were obtained using 20 realizations of Caltrans design spectrum for deep alluvium soil in seismic Zone 7. Two cases are studied (i) a tall pier (T = 0.60 sec) and (ii) a short one (T = 0.20 sec). A good correlation is observed between the simulated performances (Fig.4a,b) and the ones approximated by Eq.(11) (Fig.4c,d). The approximation was done for strength reduction damage factors D_2 of 10% which correspond only approximately to the ones calculated by simulations. This is also one of the sources of discrepancies between the simulation and the approximations.

It should be noted that the damage is more severe for piers which have low ductility, a well known fact, however using low reduction factors it is possible to keep this damage within acceptable limits for the taller columns (see Fig. 4b). It is very difficult however to control the damage in the short columns even using low reduction factors (see Fig 4a).

The energy dissipation during cyclic loading can be linked to strength deterioration as was mentioned in the previous section. Various materials respond differently to such dissipation. Steel members or highly reinforced concrete elements show a relatively small strength deterioration associated to the energy dissipation while masonry walls or low reinforced concrete elements show large strength deteriorations. The importance of the energy dissipation for deteriorating members can be observed in Fig.5. The strength reduction factors (R_D) for various damage limits are influenced by the component of strength degradation (D_2). If the serviceability limit is used for comparison, it is observed that for elements with larger strength deterioration the admissible strength reduction factors are significantly reduced. For large values of strength deterioration (D_2), repairability can be achieved only if no strength reduction factor (R_D = 1) is used.

Design of R/C Elements: Using relation in Eq.(11) design charts accounting for damage control can be developed for structural members (see Fig 6 for charts developed for highly reinforced members). It can be noted that for a targeted ductility capacity of 3, one should use a reduction factor of near 2 for a repairable short column (Fig 6a) or a reduction factor of near 3.5 for a very tall column (Fig.6b). It is clear that providing same ductility but using larger reduction factors the expected damage can be beyond repair.

DISCUSSIONS AND CONCLUSIONS

The expected seismic damage of structures has been used to control the design process. This paper defines a way to quantify the seismic damage from the deformations and the energy dissipated by structural members and suggests to use this damage as a limit criterion in establishing their strength and ductility. It is shown that the strength deterioration due to energy dissipation is very important in the selection of proper strength reduction factors.

72

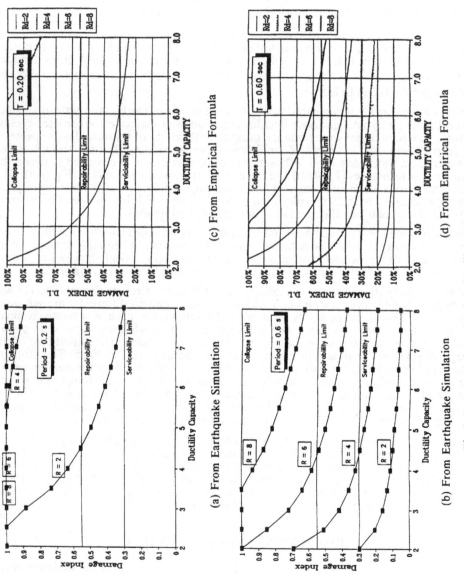

(a) From Earthquake Simulation

(b) From Earthquake Simulation

(c) From Empirical Formula

(d) From Empirical Formula

Fig.4. Damage Index related to ductility and strength reduction

73

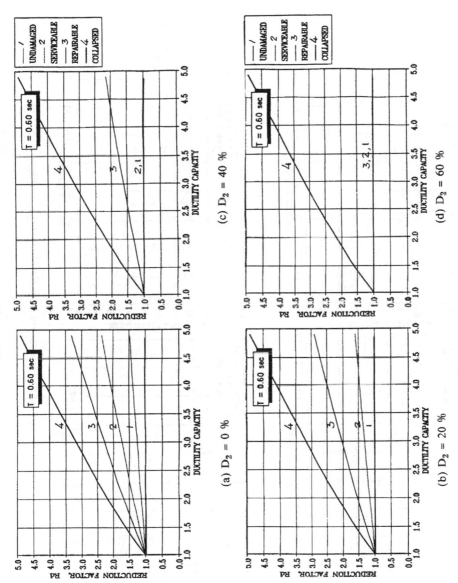

(a) $D_2 = 0$ %

(b) $D_2 = 20$ %

(c) $D_2 = 40$ %

(d) $D_2 = 60$ %

Fig.5. Influence of strength deterioration on reduction factors

74

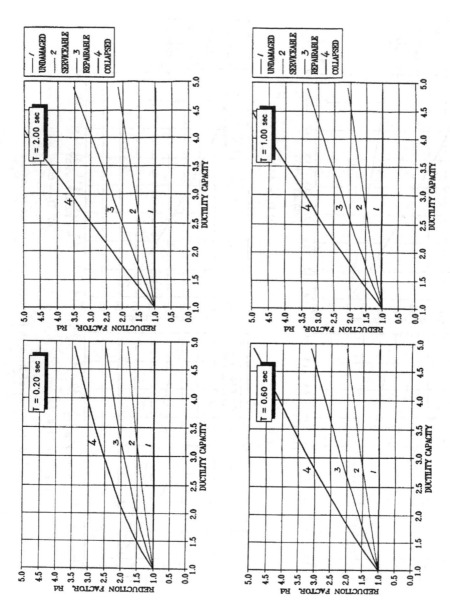

Fig.6. Design charts for strength reduction factors ($D_2 = 10\%$)

Structures prone to quick strength loss having sufficient ductility should use smaller reduction factors if low damage is desired. It was shown also that short period structures (or stiff masonry members) are extremely sensitive to the reduction factors, ie small reductions used in the design may lead to damageable structures beyond repair. Longer period structure may be designed using larger reductions.

It can be concluded that using damage states as seismic design targets, it is possible to obtain more suitable reduction factors and ductility requirements essential for reinforced concrete structures. Damage indicators that consider the energy dissipation reflected in the strength and stiffness deterioration can be used successfully to control the design parameters.

ACKNOWLEDGEMENTS

The authors wish to acknowledge support from the National Center for Earthquake Engineering Research (NCEER Grant 90-1201) which in turn is supported by the National Science Foundation (Grant No. ECE 86-07591) and the State of New York.

REFERENCES

[1]   Anagnostoupoulos, S.A., Petrovski J. and Bouwkamp, J.G. (1989), "Emergency Earthquake Damage and Usability Assessment of Buildings", *EERI/Earthquake Spectra*, Vol. 5, No. 3, August, pp 461-476

[2]   Ang, A.H.S. (1988) "Seismic Damage Assessment and Basis for Damage Limiting Design", *Probabilistic Engineering Machanics*, Vol.3, No. 3, pp. 146-150.

[3]   Banon, H., Biggs J.M. and Irvine, H.M. (1981), "Seismic Damage in Reinforced Concrete Frames", *ASCE/Journal of the Structural Division*, Vol.107, No. ST9, September, pp. 1713-1729.

[4]   Bertero, V. V. and Bressler, B. (1977), "Design and Engineering Decision: Failure Criteria (limit States), Developing Methodologies for Evaluating the Earthquake Safety of Existing Buildings", Report No. EERC-77-6, UC/Berkeley, CA.

[5]   Blejwas, T. and Bressler, B. (1979), "Damageability of Existing Buildings", Report No. EERC-78-12, UC/Berkeley, CA.

[6]   Bracci, J. M., Reinhorn, A. M., Mander, J. B. and Kunnath, S. K.(1989), "Deterministic Model for Seismic Damage Evaluation of Reinforced Concrete Structures", Technical Report #NCEER-89-0033, National Center for Earthquake Engineering Research - SUNY/Buffalo, NY.

[7]   Chung, Y. S., Meyer, C. and Shinozuka, M. (1987), "Seismic Damage Assessment of Reinforced Concrete Members", Technical Report#NCEER-87-0022, National Center for Earthquake Engineering Research - SUNY/Buffalo.

[8]   Darwin, D. and Nmai, C.K. (1986), "Energy Dissipation in RC Beams under Cyclic Load", *ASCE/Journal of Structural Engineering*, Vol.112, No. 8, August, pp. 1826-1846.

[9]   DiPasquale, E. and Cakmak, A.S. (1989), "On the Relation Between Local and Global Damage Indices", Technical Report #NCEER-89-0034, National Center for Earthquake Engineering Research - SUNY/Buffalo, N.Y.

[10]   Gosain, N.K. Brown, R.H. and Jirsa, J.O. (1977), "Shear Requirements for Load Reversals on RC Members", *ASCE/Journal of the Structural Division*, Vol.103, No. ST7, July, pp. 1461-1476.

[11] Hidalgo, P.A. and Arias A. (1990), "New Chilean Code for the Earthquake Resistant Design of Buildings", Proceedings of Fourth U.S. National Conference on Earthquake Engineering, Vol.2, Palm Springs Ca., pp. 927-936.

[12] Hwang, T.H. and Scribner, C.F. (1984), "R/C Member Cyclic Response During Various Loadings", *ASCE/Journal of Structural Engineering*, Vol.110, No. 3, March, pp. 477-489.

[13] Krawinkler, H. and Zohrei, M. (1983), "Cumulative Damage in Steel Structures Subjected to Earthquake Ground Motion", *Computers and Structures*, Vol 16, No.14.

[14] Lybas, J. and Sozen, M.A. (1977), "Effect of Beam Strength and Stiffness on Dynamic Behavior of Reinforced Concrete Coupled Walls", Civil Engineering Studies, SRS No. 444, University of Illinois, Urbana, IL.

[15] Mander, J.B., Priestley, M.J.N. and Park, R. (1983), "Behavior of Ductile Hollow Reinforced Concrete Columns", *Bulletin of the New Zealand National Society For Earthquake Engineering*, Vol. 16, No. 4, December, pp.273-290.

[16] McCabe, S.L. and Hall W.J., (1989), "Assessment of Seismic Structural Damage", *ASCE/Journal of Structural Engineering*, Vol.115, No. 9, August, pp. 2166-2183.

[17] Meskouris, K. and Kratzig, W.B. (1987), "Nonlinear Seismic Response of Reinforced Concrete Frames", Proceed. of Nonlinear Stochastic Dynamic Engineering Systems, IUTAM Symposium, Innsbruck, Austria.

[18] Osteraas, J. and Krawinkler, H. (1990), "Seismic Design Based on Strength of Structures", Proceedings of Fourth U.S. National Conference on Earthquake Engineering, Vol.2, Palm Springs Ca., pp. 955-964.

[19] Park, Y.J. Ang, A.H-S. and Wen, Y.K. (1985), "Seismic Damage Analysis of Reinforced Concrete Buildings", *ASCE/Journal of the Structural Engineering*, Vol.111, No. 4, April, pp. 740-757.

[20] Ridell, R., Hidalgo P. and Cruz, E., (1989), "Response Modification Factors for Earthquake Resistant Design of Short Period Buildings", *Earthquake Spectra*, Vol.5, No.3, August, pp. 571-590.

[21] Roufaiel, M.S.L. and Meyer, C. (1985), "Analysis of Damaged Concrete Frame Buildings", Report No. NSF-CEE81-21359-1, Dept. of Civil Engineering, Columbia University, N.Y.

SEISMIC ANALYSIS OF DEGRADING MODELS
BY MEANS OF
DAMAGE FUNCTIONS CONCEPT

EDOARDO COSENZA
Istituto di Ingegneria Civile, Università di Salerno
Penta di Fisciano, 84080, Salerno, Italy

GAETANO MANFREDI
Istituto di Tecnica delle Costruzioni, Università Federico II
P.le V. Tecchio, 80125, Napoli, ITALY

ABSTRACT

The seismic verification of structures subjected to high intensity ground motions requires on the one hand the definition of a suitable deterioration model of the structure, and on the other hand the definition of a damage function; structural collapse is expected if this function becomes greater than 1 during the seismic event.

In this paper a review of the models used in non linear seismic analysis is presented; in particular the classic elastic-perfectly-plastic model is considered as reference-models, and the other models are classified as non-evolutionary and evolutionary, and degrading and non-degrading. Thus, some of the function previously developed are critically analyzed, with particular reference to the maximum ductility, cumulative dissipated energy, their possible combinations and linear cumulative damage theory function.

The response spectra in terms of acceleration are given, considering some significant acceleration records, and a comparison between the results relative to the different evolutionary models and definition of the seismic failure is provided.

Finally, first results about a new evolutionary-degrading general model are provided.

INTRODUCTION

The verification of structures subjected to high intensity ground motion can be performed by means of the following steps: (1) definition of a model that actually represents the cyclic behaviour of structure under examination; (2) definition of a "Damage Function" that actually represents the cyclic behaviour of the structure; (3) evaluation of the values of the function that are "compatible" with the safety of the structure, in relation to the materials and to the structural system analyzed; (4) evaluation, with reference to the local site conditions and suitable attenuation laws, of the expected destructive earthquake, and, by means of theoretical nonlinear dynamic analysis, of the values of the function "required" to avoid the failure. The seismic check is, hence, performed by means of the comparison between the "compatible" and the "required" values of the damage function; in the case of "Normalized Functions" the required value of the damage function must be less than unity.

With reference to the choice of the structural model, we can observe that models that are completely indifferent to the evolution of the strain cycles and to the structural damage exist; for example, elastic-perfectly plastic model belongs to this class.

Other models exhibit unloading and/or reloading rules depending on the maximum values of the previously deformations; this class is represented by models like Clough-Johnston [11], Slip, Q-model [37], origin oriented model [16,29,36] etc..

An other class is represented by models that degrade with cycles; the most widely used seem to be the Wen model [3,4,32,43,44]; very interesting general model is also given in [25], and, with specific reference to r.c. sections, in [8,9,10,20,28] and in [35].

With reference to the choice of an optimum damage function described at point (2), it is necessary to observe that different levels of damage function definitions exist.

The simplest definition requires the use of a unique global parameter; the most widely used is relative the "maximum ductility", i.e. the maximum amount of the plastic displacement, adequately non-dimensionalized. Another choice of global parameter is obtained considering the cumulative dissipated energy or the input energy [1,2,17,40,42]. These global approaches, that have the merit of simplicity, may be subjected to criticism, because they completely neglect the real loading history. In other words, the ductility criteria considers only the maximum displacement cycle, without any information with regard to the other cycles; at the opposite side the dissipated energy criteria considers the total amount of the plastic dissipation, independently by the cumulative modality. From the experimental point of view, we can affirm that both the hypothesis are often far from the reality.

More realistic results may be obtained considering the combination, linear or non-linear, of these couple of damage parameters; for this scope some authors suggest the choice of a damage function which considers both the maximum inelastic displacement and the dissipated energy [6,7,30,31], by means of formulation tried by experimental data.

Very effective is considering, even if in approximate way, the distribution of plastic cycles [12]; some authors consider this distribution of the inelastic cycles by means of the linear cumulative damage theory [23,24,26,34,45,46]. In these formulations, the damages due to different plastic deformations are accumulated, using an adequate weighting function; in this way, a damage function which considers the effective distribution of the plastic cycles is defined, although the exact order of the plastic deformations is not considered; the weighting function provides a major importance to the greatest deformations, and a minor importance to the lowest deformations. As first limit case, if all the plastic deformation have the same importance, the damage function gives the same results of the energy one; in the opposite limit case, only the maximum deformation cycle is considered and the function provides the same results of the ductility one.

A more sophisticated damage function can be provided considering the sequence of the different plastic rates [8,9,10]; in this way it is possible to obtain an interesting extension of the linear cumulative damage theory.

In all the definitions introduced, it is postulated that different loading histories that lead to the same damage function, exhibit the same structural deterioration. Therefore, with reference to the point (3), the limit value of the damage function can be evaluated by means a monotonic test brought up to failure. With reference to the point (4), furthermore, the analysis can be performed evaluating the acceleration that leads the damage function to the unity, and that, hence, leads the structural model to the seismic collapse. It is than possible to define collapse spectra associated to the given definition of damage function.

THE CLASSIFICATION OF THE STRUCTURAL MODELS
UNDER CYCLIC LOADING

Structural models under cyclic loading are characterized by the following definitions:
1) definition of monotonic envelop; the envelop under cyclic loading is generally assumed to coincide with the monotonic envelop;
2) definition of the unloading rules; in the simplest case we assume that the unloading is linear;
3) definition of the reloading rules; the reloading may follow complicate rules and can generate slip, pinching, etc.

By means of the specification of these definition we can consider the following classes of models:

      Ia) non-evolutionary models;
      Ib) evolutionary models:

and:

      IIa) non-degrading models;
      IIb) degrading models.

The evolutionary models are characterized by reloading depending on the actual strain; in other words the reloading rules depends on the status (but not on the damage) of the system. Example of non-evolutionary model are the elastic-perfectly-plastic (EPP) and elasto-plastic with hardening (EPH) or softening (EPS) models. Evolutionary models are the origin oriented model [16,29,36], Clough-Johnston model [11], Slip model, Q-model [37] etc. It should be noted that EPP, EPH and EPS, from a theoretical point of view, are completely different from the others; in fact they are characterized by unloading and reloading that are elastic or, in other words, without residual plastic deformations; as a consequence the plasticity of these models is active if and only if the plastic bound (ultimate strength) is reached. In the other models residual plastic deformations may occur also on the reloading branch, without to reach the plastic bound: as a consequence these models are completely out from the classic plasticity theory frame.

The degrading models are characterized by unloading and reloading rules depending on the damage state of the system; in other words a non-degrading model subject to equal amplitude strain cycles provides always the same cycle; on the contrary a degrading model subject to equal amplitude strain cycles provides a behaviour that progressively degrade, as a function of the damage function which governs the problem, up to the failure. Example of this couple of different behaviours is given in Fig.1.

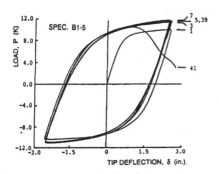
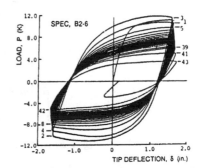

Steel element with large deterioration   Steel element with gradual deterioration
(crack propagation)   (local buckling)

Fig. 1 Failure models with different deterioration behaviour [24].

It seems interesting to observe that all the evolutionary and non-evolutionary models indicated previously may be both degrading or non-degrading; for example, the case of elastic perfectly-plastic model with degrading unloading provides the model analyzed in the damage continuum mechanics, where plasticity is always uncoupled from elasticity, but where elasticity continuously degrades.

Even if the experimental behaviour of structure is often of degrading type, with rules that evolve with the progressive damage, the evolutionary degrading models are scarcely used in the technical literature; some of these are presented in [3,4,32,43,44] and [25]. Both models provide a deterioration related to the plastic dissipated energy; in particular in the second model the parameter β (see next section) plays an important rule. More general models for r.c. structures are presented in [8,9,10] and [35].

The analysis of seismic behaviour of evolutionary models plays a very important role for the complete understanding of the seismic behaviour of structures; on the other hand, in a general context, the deterioration of the model has to be related to the general concept of damage function and not necessarily to the plastic dissipated energy. For this reason final sections of this paper will be devoted to the development of a general evolutionary model that, as particular case, may provide non-degrading models.

THE DEFINITION OF THE DAMAGE FUNCTIONS

In the following the term "Damage Function" will be used to define a function that is defined on the space of parameter considered of interest in the cyclic-seismic phenomena (maximum deformation, plastic dissipated energy, etc.) and with value in the real field; in the case of "Normalized Damage Functions" it assumes the value 0 in the case of absence of structural damage and the value 1 in the case of failure.

Among the different damage functions defined in technical literature we can use the following classification:

1) one seismic parameter functions: ductility (monotonic or cyclic) method or plastic dissipated energy method.
2) two seismic parameters functions: ductility plus plastic energy;
3) low cycle fatigue methods.

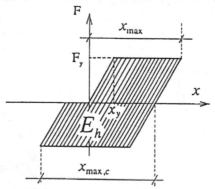

Fig. 2 Elastic-Perfectly Plastic (EPP) model.

Class 1) of damage functions is represented by ductility or dissipated energy; in particular the most widely used structural damage parameter is the *cinematic ductility* (cfr. Fig.1):

$$\mu_s = \frac{x_{max}}{x_y} \tag{1}$$

where x_{max} denotes the maximum plastic displacement of the structural model, and x_y the yielding displacement.

Due to the frequent variations of the signs of seismic acceleration, the response is characterized by plastic cycles with different sign displacements; for this reason the maximum (positive) or the minimum (negative) value of the displacement, in some cases, does not seem to be the most realistic measure of the plastic deformation of the model. To quantify, hence, the maximum plastic engagement, it is convenient to define the *cyclic ductility* [27] in the following way (Fig.1):

$$\mu_c = \frac{x_{max,c}}{x_y} \tag{2}$$

where $x_{max,c}$ denotes the maximum plastic excursion independently from the sign.

A well defined damage function D_μ can be immediately associated to the definitions of cinematic or cyclic ductility; defining the allowable value of the ductility $\mu_{u,mon}=x_{u,mon}/x_y$ by means of a monotonic test, the damage function may be defined as:

$$D_\mu = \frac{\mu - 1}{\mu_{u,mon} - 1} \tag{3}$$

The choice of the cinematic or cyclic ductility as damage measure is equivalent to assume that the collapse of the structural model is expected for maximum plastic displacement, independently from the cycles number and the amount of dissipated energy. Experimental tests have shown that this type of behaviour may represent satisfactorily the structural deterioration in the case of cycles characterized by one cycle with large plastic displacement, and the others with a little amount of plastic engagements.

If the loading history is characterized by many cycles with large amount of plastic deformation, or the structure has a little attitude to dissipate energy, the failure depends, beside the maximum displacement, on the total amount of plastic dissipated energy E_h. This energy is normalized, thus becoming comparable to the ductility, by the definition of *hysteretic ductility* [27]:

$$\mu_e = \frac{E_h}{F_y x_y} + 1 \tag{4}$$

where F_y is the strength of the structural model. The value of the corresponding normalized damage function D_E may be defined, analogously to the cinematic and cyclic ductility, evaluating experimentally the allowable hysteretic ductility $\mu_{e,u,mon}$ in a monotonic test. This value, in the elasto-plastic model, is equal to $\mu_{u,mon}$. D_E has hence the following expression:

$$D_E = \frac{\mu_e - 1}{\mu_{e,u,mon} - 1} = \frac{\mu_e - 1}{\mu_{u,mon} - 1} \tag{5}$$

Class 2) provides more realistic damage functions, which consider in the deterioration both the effect of the plastic displacement and of the plastic dissipated energy; in particular, with reference to the reinforced concrete elements, with combined axial load and bending, Banon et al. [6,7], and Park & Ang [30,31] consider this type of functions that have also the great advantage to be defined by means of one combination parameter that may be calibrated by experimental results.

In the first case the two damage parameters D_1 and D_2 are defined, respectively as the ratio of the stiffness at the yielding point to the stiffness at failure, and the plastic dissipated energy normalized respect to the elastic one; in the elastic-perfectly plastic model the value of D_1 is obviously equal to the ratio of the maximum displacement to the yielding displacement. Thus, with the symbolism previously introduced, we have:

$$D_1 = \frac{x_{max}}{x_y} = \mu_s \quad ; \quad D_2 = \frac{E_h}{1/2 F_y x_y} = 2(\mu_e - 1) \tag{6}$$

Furthermore, defining the modified damage parameters D_1^* and D_2^* respectively as D_1-1 and aD_2^b, where the average experimental values of a is 1.1, and the choice $b=0.38$ is suggested [6], the following damage function D_B is introduced:

$$D_B = \sqrt{(D_1^*)^2 + (D_2^*)^2} \quad with \quad D_1^* = (\mu_s - 1) \quad ; \quad D_2^* = 1.1[2(\mu_e - 1)]^{0.38} \tag{7}$$

The corresponding normalized function is obtained dividing the expression (7) by the value that assumes at failure in a monotonic test.

The Park & Ang damage function is defined as the linear combination of the maximum displacement and the dissipated energy, in the following way:

$$D_{P.A.} = \frac{x_{max}}{x_{u,mon}} + \beta \frac{E_h}{F_y x_{u,mon}} = \frac{\mu_s + \beta(\mu_e - 1)}{\mu_{u,mon}} \tag{8}$$

where the parameter β depends on the value of shear and axial forces in the section and on the total amount of longitudinal and confining reinforcement, by means of a regression curve obtained from more than 250 experimental results. The experimental values of β reported in [30] ranged between about -0.3 to +1.2, with a median of about 0.15; in [33] the choice β=0.025 for steel structures and β=0.05 for concrete structures is suggested. The function exhibits the merit, beside to take into account of both maximum plastic displacement and plastic dissipated energy, to consider the structural deterioration as a function of the structural characteristics, by means of the parameter β. At last, the expression (8) does not provide the value 0 for $x_{max}=x_y$, and does not provide the value 1 in the case of failure under monotonic loading.

The methods of Banon et al. and Park & Ang do not take into account the plastic cycles distribution. In other words, they do not consider the modality of achievement of the total amount of dissipated energy.

The class 3) of damage functions includes the possibility to consider, even if in a simplified way, the distribution of plastic cycles; in particular a damage function D_F which takes into account the different amount of the plastic displacements can be defined [23,24] using the following law of the cumulative damage:

$$D_F = A \sum_{i=1}^{n} (\mu_i - 1)^b \tag{9}$$

where A and b are structural constants, n is the total number of plastic cycles, and μ_i is the ductility (cinematic or cyclic) relative to the ith plastic cycle. In particular, a monotonic test provides:

$$1 = A(\mu_{u,mon} - 1)^b \quad \Rightarrow \quad A = \frac{1}{(\mu_{u,mon} - 1)^b} \tag{10}$$

Thus, it is immediate to obtain the following result:

$$D_F = \sum_{i=1}^{n} \left(\frac{\mu_i - 1}{\mu_{u,mon} - 1} \right)^b \tag{11}$$

The value assumed by this damage function is defined by means of the constant b, which depends on the structural material and system, and on the amount of the different plastic displacements, independently from the exact order. Typical values of b, obtained by experimental data for steel structures [23,34] and reinforced concrete structures [34,38,39], are 1.6 to 1.8; in the damage analysis, some times, the conservative value 1.5 is assumed [6].

The value b=1 in eqn. (11) gives the same weight to each plastic displacement, independently from its amount; hence, the function D_F reaches the unity value at the same time of the energy function D_E. On the other side, if b assumes very large values, eqn. (11) gives a always decreasing importance to the smaller plastic displacements, and the function D_F reaches the unity at the same time of D_μ; thus, we can affirm that D_F provides values always included between D_μ and D_E. It is also interesting to observe that the cumulative damage law (11) admits also the well known form of the linear cumulative law of fatigue. The validity of (13) to analyze low cycle fatigue in steel components is shown in [23,45,46], and in reinforced concrete section in [38,39]. Best results for r.c. structures may be obtained by means of the interesting generalization of eqn. (13) presented in [8,9,10], which allows to consider the order of application of plastic cycles, non symmetric behaviour etc.

SEISMIC RESPONSE OF THE EPP MODEL

The analysis of seismic behaviour of the reference model EPP (non-evolutionary, non-degrading) has been extensively provided in [14,15]; in the following main results will be summarized.

In particular, to examine some properties of Damage Functions, consider the recorded ground motions of Rocca (Ancona, Central Italy, 1972), Tolmezzo (Friuli, Northern Italy, 1976), Calitri (Irpinia, Southern Italy, 1980), and Mexico City (1985); these records were selected in order to analyze very different types of earthquakes. In fact, in the first case, spectra with amplification only for low periods are obtained; in the second case the amplification is relative to the medium range periods, and in the fourth case it is relative to very high periods; in the case the Calitri the amplifications concern to a large range of periods. The synthetical characterization of these registration and the correction filters used in the analysis are presented in tab. 1.

SEISMIC EVENT	DATE	RECORD	COMP.	low p. filter [Hz]	high p. filter [Hz]	a_g [g]	t [s]	t_D [s]	I_A [m/s]	I_v [m,s]
Ancona	14.06.72	Rocca	NS	.10-.33	25-27	.603	19.0	2.55	.693	.155
Friuli	06.05.76	Tolmezzo	WE	.10-.33	25-27	.313	36.4	4.92	.786	.434
Campano-Lucano	23.11.80	Calitri	WE	.10-.33	25-27	.166	86.1	47.5	1.376	.744
Ixtapa, Mexico	19.09.85	Mexico City	NW	.05-.055	23-25	.171	180	38.9	2.541	1.511

Tab. 1. Recorded ground motions considered in the analysis. a_g= max. ground acceleration; t=registration duration; t_D=duration after Trifunac and Brady; I_A= intensity after Arias; I_v= intensity after Fajfar et al. [18,19].

In [14,15], considering a structure characterized by ultimate monotonic ductility $\mu_{u,mon}$=4, the collapse acceleration spectra of the EPP are obtained; the seismic failure is reached when D=1, and different definition of damage function are used. Some results are presented in Figs. 3 and 4.

In particular in all the figures the spectra considering the ductility and the energy methods are given; moreover the figures provide:
Fig. 3: design spectra for Tolmezzo (left side) and Mexico City (right side) records considering:
- Banon and Veneziano criteria, with mean suggested value of a=1.1 and for the value of a that practically provides the energy method (a=2);
- Park and Ang criteria, with mean suggested values of β=0.05 and 0.015 and for the values of β that practically provides the energy method (β=0.6÷0.8);
- Krawinkler criteria, with suggested values of b=1.5, 1.8, 2.0 and for the values of b that practically provides the energy method (b=5÷10).
Fig. 4: design spectra for all the records of Tab.1 with the comparison among Banon and Veneziano, Park and Ang, Krawinkler criteria considering the mean values a=1.1, β=0.15 and b=1.8.

By means of these figures the following conclusions [14,15] may be summarized:
1) the scatter of results obtained considering the ductility or the plastic dissipated energy is very large.
2) Banon and Veneziano damage function with a=2 and Park and Ang damage function with β=0.6÷0.8 practically provide the same result of energy method; greater values of these parameter does not seem have physical significance.
3) Krawinkler damage function with b>5 practically provides the same result of ductility method; with greater values of this parameter it seems logical to consider the simple ductility function method.
4) Banon & Veneziano, Park & Ang and Krawinkler damage functions, with suggested values of the parameters (a=1.1, β=0.15 and b=1.8), provide results practically coincident.

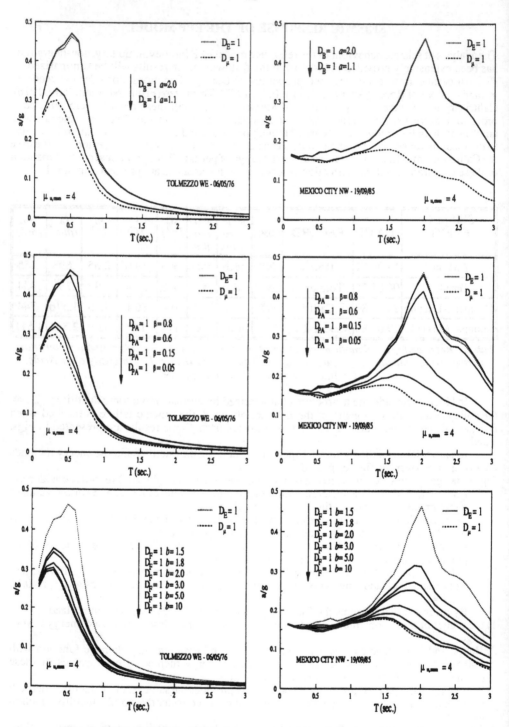

Fig. 3. Collapse acceleration spectra using different Damage Functions.

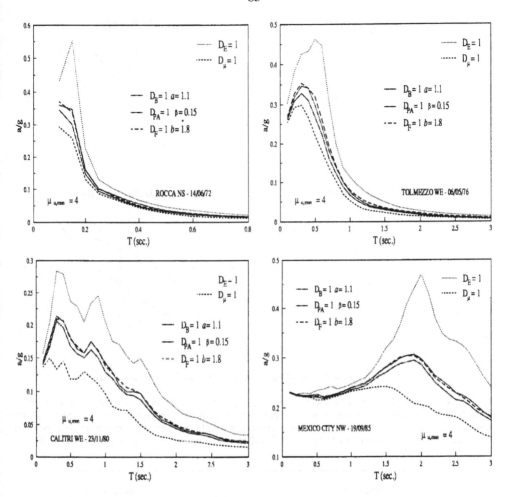

Fig. 4. Comparison between different Damage Functions with mean values of parameters.

SEISMIC RESPONSE OF EVOLUTIONARY NON-DEGRADING MODELS

In this section the response of the Clough-Johnston, Origin oriented and Slip model are compared to the response of the EPP model (cfr. Fig. 5); all the damage functions previously introduced are considered in the analysis.

The first comparison among the models is presented in Fig. 6, where acceleration spectra considering the cinematic ductility and the plastic dissipated energy damage functions are drawn, with reference to the Tolmezzo record. In the evaluation of the spectra the value $\mu_{u,mon}$=4 is considered; the limit plastic energy is evaluate with reference to the standard cycles with maximum plastic deformation; therefore, the value of $\mu_{e,mon}$ for the Clough-Johnston and slip models is assumed equal to $1/2\,\mu_{u,mon}$ and for the origin-oriented model to $1/4\mu_{u,mon}$.

The comparison between the spectra allows to observe that both ductility and energy spectra increase; the increase of the ductility controlled spectra is very little and the increase of the hysteretic controlled spectra is higher: therefore the distance between the criteria increases. In other words the behaviour of structures governed by the ductility damage function seems to be only a little influenced by the modelling; on the contrary the behaviour of the structures controlled by the energy damage function is more sensible to the modeling.

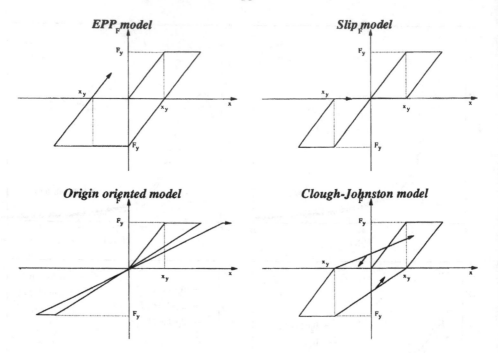

Fig.5. Reference and evolutionary models considered in the analysis.

In Fig. 7 the acceleration spectra considering also Park & Ang and Krawinkler damage functions are drawn; in the comparison between different damage functions it is interesting to observe that slip and origin-oriented models provide results very close to ductility method. This fact is due to the little number of plastic cycles that characterizes the seismic behaviour of these models.

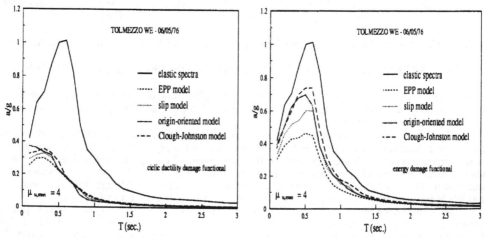

Fig. 6. Comparison between ductility and dissipated energy damage functions spectra for evolutionary non-degrading models.

An other interesting observation is that to have the same results with Park & Ang and Krawinkler damage functions the value β=0.4 has to be chosen in the Clough and origin-oriented model; in fact these models provide a lower dissipation of energy and, thus, a larger value of the coefficient has to be considered. This result is in agreement with the trend of the parameter β given in [30], where r.c. structures characterized by low deterioration capacity (high shear, high axial force, low confining reinforcement, etc.) have higher value of β.

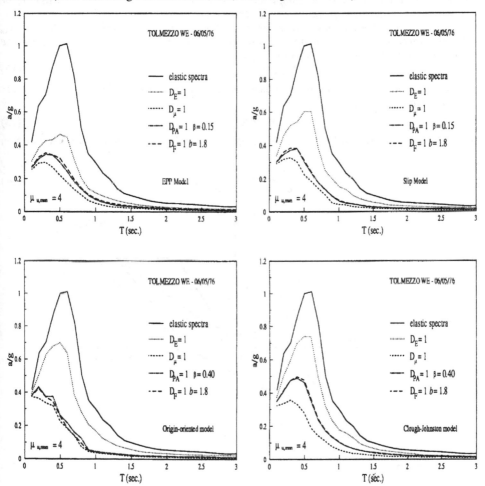

Fig. 7. Comparison between different Damage Functions with mean values of parameters for evolutionary models.

A PROPOSAL OF GENERAL EVOLUTIONARY-DEGRADING MODEL

Among the general degrading models, the Wen model [3,4,32,43,44] and Kunnath et al. model [25] seem to be the most attractive. The first exhibits the interesting property, especially for sthocastic analysis, to be defined by means of differential equations; some interesting observation on this model are presented in [41]. The second model is strictly connected with the Park & Ang criteria and provides an accurate physical interpretation of the parameter β of eqn.

88

(8); furthermore it may be used to analyze different structural elements (beams, columns, shear-walls, etc) and gives as particular cases the previously introduced non degrading models. Both models provide degrade is caused by the plastic dissipated energy.

A general evolutionary-degrading model [13], that may be associated to all the Damage Functions previously defined, may be developed considering the following rules:

1) *the monotonic envelop may be completely generic, with perfect plasticity, hardening or softening; the cyclic envelop coincides with the monotonic one;*

2) *considering, for simplicity, unloading and reloading of linear type, unloading stiffness K_{un} must be always not greater than tangent initial stiffness K_o and reloading stiffness K_r must be always not greater than unloading stiffness K_{un}:*

$$K_{un} \leq K_o \quad ; \quad K_r \leq K_{un} \tag{12}$$

3) *it is not possible to have unloading stiffness K_r less than the minimum stiffness of the monotonic envelop, that is $K_o/\mu_{u,mon}$:*

$$K_r \geq K_o/\mu_{u,mon} \tag{13}$$

4) *the unloading stiffness and the reloading stiffness must decrease with the damage function D:*

$$\frac{\partial K_{un}}{\partial D} \leq 0 \quad ; \quad \frac{\partial K_r}{\partial D} \leq 0 \tag{14}$$

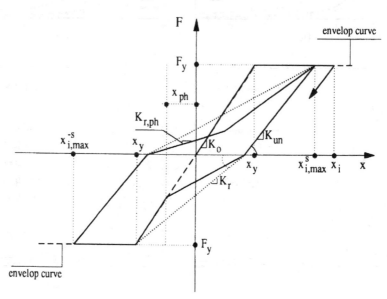

Fig. 8. Definition of parameters of the proposed evolutionary-degrading model.

Synthetically, defining the functions F_{un} and F_r as the ratio between unloading stiffness and tangent stiffness and between reloading stiffness and unloading stiffness:

$$F_{un} = \frac{K_{un}}{K_o} \quad ; \quad F_r = \frac{K_r}{K_{un}} \tag{15}$$

these functions must have the following properties:

$$0 \leq F_{un} \leq 1 \quad ; \quad \frac{1}{\mu_{u,mon}} \leq F_r \leq 1 \tag{16}$$

$$\frac{\partial F_{un}}{\partial D} \le 0 \quad ; \quad \frac{\partial F_r}{\partial D} \le 0 \tag{17}$$

Pinching on the reloading branch may be activated by means of the introduction of the following non-dimensional parameters (cfr. Fig. 8):

$$F_{ph} = \frac{K_{r,ph}}{K_r} \le 1 \quad ; \quad \mu_{ph} = \frac{x_{ph}}{x_y} \le 1 \tag{18}$$

Among the infinite functions that provides the above mentioned properties, the following simple functions will be analyzed:

$$F_{un} = \frac{1}{1 + D^{\alpha}(\mu^s_{i,max} - 1)} \tag{19}$$

$$F_r = \frac{1}{1 + \beta[(|\mu_i| + \mu^{-s}_{i,max})(1 + \gamma D^{\delta}) - 1]} \tag{20}$$

where:

μ_i = ductility of the last plastic excursion;

$\mu^s_{i,max}$ = max. ductility among the previous cycles with sign equal to μ_i:

$$\mu^s_{i,max} = \max_{j \le i} \left\{ \frac{sgn\mu_j + sgn\mu_i}{2} \mu_j \right\} \tag{21}$$

$\mu^{-s}_{i,max}$ = max. ductility among the previous cycles with sign opposite to μ_i:

$$\mu^{-s}_{i,max} = \max_{j \le i} \left\{ \frac{sgn\mu_j - sgn\mu_i}{2} \mu_j \right\} \tag{22}$$

The parameter α defines the degrade of unloading stiffness and β defines the degrade of the reloading stiffness; γ defines the maximum cyclic degrade of reloading stiffness and δ the velocity of this degrade. In particular, considering combination of the parameters given in Tab. 2, we have all the non-degrading models analyzed previously.

	α	β	F_{ph}, μ_{ph}	γ
EPP	∞	0	1, inactive	inactive
Origin oriented	0	0	1, inactive	inactive
Slip	∞	0	0, 0	inactive
Clough-Johnston	∞	1	1, inactive	0

Tab. 2. Non evolutionary models as particular cases of the general model.

It seems interesting to observe that using a finite value of α and $\beta=0$, $F_{ph}=1$ we have the mono-dimensional model used in damage continuum mechanics [22].

It seems also interesting to observe that, considering α infinite, we have a class of Clough-type models with reloading less stiff than classic Clough-Johnston model for $\beta<1$ and more stiff for $\beta>1$; cyclic degrade of the model is active for $\gamma>0$.

To better understand the physical significance of the parameters γ and δ consider the classic Clough-Johnston model ($\alpha = \infty$, $\beta = 1$); eqns. (19)-(20) immediately provide:

$$\left(\frac{1}{K_r}\right)_D = \left(\frac{1}{K_r}\right)_{D=0} (1 + \gamma D^{\delta}) \tag{23}$$

and then, at failure:

$$\left(\frac{1}{K_r}\right)_{D=1} = \left(\frac{1}{K_r}\right)_{D=0} (1 + \gamma) \tag{24}$$

For equal amplitude cycles, thus, the reloading stiffness K_r degrades, with velocity governed by δ, from an initial value to the limit value $1/(1+\gamma)$ times lower; therefore also a lower bound cyclic envelop exists, and this limit envelop may be easily provided by the monotonic envelop and the parameter γ. A similar properties characterizes the model [35].

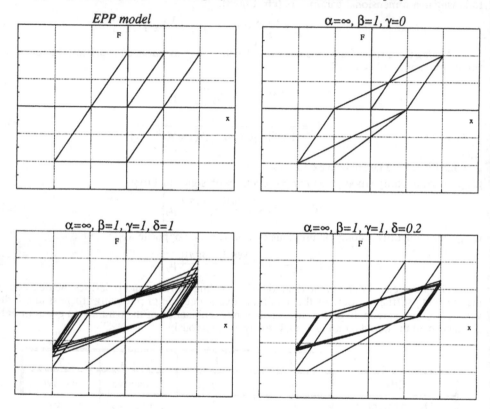

Fig. 9. Degrading models considered in the analysis.

FIRST RESULTS ABOUT SEISMIC RESPONSE
OF PROPOSED EVOLUTIONARY-DEGRADING MODEL

In this section some preliminary results relative to the proposed evolutionary-degrading model will be presented. In particular consider the following modeling (cfr. Fig. 9):
   1) reference EPP model ($\alpha=\infty$, $\beta=0$);
   2) classic Clough-Johnston model ($\alpha=\infty$, $\beta=1$, $\gamma=0$);
   3) Clough-Johnston model with reloading degrade activated and linear reloading degrade velocity ($\alpha=\infty$, $\beta=1$, $\gamma=1$, $\delta=1$);
   4) Clough-Johnston model with reloading degrade activated and accelerated reloading degrade velocity ($\alpha=\infty$, $\beta=1$, $\gamma=1$, $\delta=0.2$).
   In Figs. 10 the acceleration spectra considering cinematic ductility, cyclic ductility, dissipated energy and plastic fatigue damage functions are drawn. In each figure the spectra considering the four considered models are given, together with the elastic spectra. A value of $\mu_{u,mon}=4$ is considered; in the case of the energy method the value of $E_{h,u}$ relative to the failure cycle with $\mu_{u,mon}=4$ is considered for the non-dimensionalization.

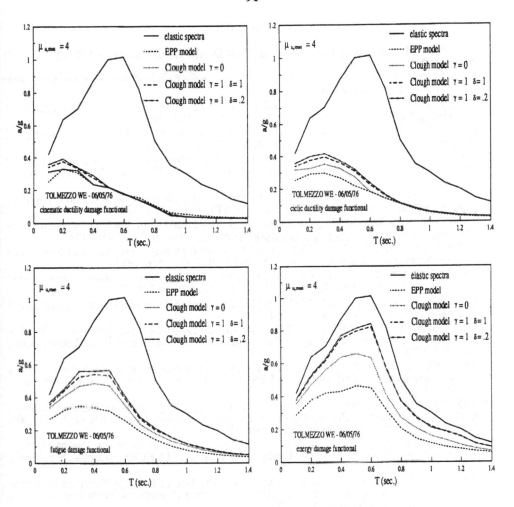

*Fig. 10. Comparison among different degrading models
using different damage functions.*

It is immediate to observe that in the case of cinematic ductility the choice of the model has only a little importance; this fact may be explained considering that the peak response generally occur at the beginning of the earthquake, when the reloading is only a little, or nothing, degraded. This result has a very important technical consequence; in fact it allows that if the seismic behaviour of a structure is governed by the ductility, then a good prevision of the behaviour may be developed by means of the classic EPP and considering sophisticated modeling is not necessary.

On the contrary, the importance of modeling increases considering cinematic ductility, because in this case the entire plastic cycle is important, and, thus, the reloading mechanism plays a relevant role; the reloading mechanism affects also the fatigue behaviour, due to the unloading that may occur on this branch, and in a greater way the dissipated energy.

In this last case the spectra with degrading model is very close to the elastic spectra; this fact means that degrading structures with associated the energy method could not try practical vantage from their ductility.

Of course, these very important results should be confirmed by more extended analyses.

REFERENCES

1. AKIYAMA H., Earthquake Resistant Limit-State Design for Buildings, University of Tokyo Press, 1985.
2. AKIYAMA H., Earthquake Resistant Design Based on the Energy Concept, paper 8-1-2, Vol. V, IX WCEE, Tokyo-Kyoto, August 2-9, 1988.
3. BABER T.T., NOORI M.N., Random Vibration of Degrading Pinching Systems, Journal of Engineering Mechanics, ASCE, Vol. 111, No. 8, 1010-1026, 1985.
4. BABER T.T., WEN Y.-K., Random Vibration of Hysteretic, Degrading Systems, Journal of Engineering Mechanics, ASCE, Vol. 107, No. EM6, 1069-1087, 1981.
5. BAIK S.W., LEE D.G., KRAWINKLER H., A Simplified Model for Seismic Response Prediction of Steel Frame Structures, paper 7-6-4, Vol. V, IX WCEE, Tokyo-Kyoto, August 2-9, 1988.
6. BANON H., BIGGS J., IRVINE H., Seismic Damage in Reinforced Concrete Frames, Journal of the Structural Division, ASCE, Vol. 107, No. ST9, 1713-1728, 1981.
7. BANON H., VENEZIANO D., Seismic Safety of Reinforced Concrete Members and Structures, Earthquake Engineering and Structural Dynamics, Vol.10, 179-193, 1982.
8. CHUNG Y.S., MEYER C., SHINOZUKA M., Seismic Damage Assesment of Reinforced Concrete Members, Technical Report NCEER-87-0022, October, 1987.
9. CHUNG Y.S., MEYER C., SHINOZUKA M., A New Damage Model for Reinforced Concrete Structures, 12-1-5, Vol. VII, IX WCEE, Tokyo-Kyoto, August 2-9, 1988.
10. CHUNG Y.S., MEYER C., SHINOZUKA M., Modeling of Concrete Damage, ACI Structural Journal, Vol.86, n.3, May-June, 259-271, 1989.
11. CLOUGH R.W., JOHNSTON S.B., Effect of Stiffness Degradation on Earthquake Ductility Requirements, Proceedings of Japan Earthquake Engineering Symposium, Tokyo, Japan, 227-232, 1966.
12. COSENZA E., MANFREDI G., Low Cycle Fatigue: Characterization of the Plastic Cycles due to Earthquake Ground Motion, RILEM Workshop on "Needs in Testing Materials", Naples, May 1990.
13. COSENZA E., MANFREDI G., An Evolutionary-Degrading Model for Damage Analysis Under Seismic Actions, Proceedings of Fifth Italian National Conference on Earthquake Engineering (in italian), Palermo, September, 1991.
14. COSENZA E., MANFREDI G., RAMASCO R., An Evaluation of the Use of Damage Functionals in Earthquake-Resistant Design, Proceedings of IX ECEE, Moscow, September, Vol. 9, 303-312, 1990.
15. COSENZA E., MANFREDI G., RAMASCO R., The Use of Damage Functions in Earthquake-Resistant Design: a Comparison Among Different Procedures, Technical Report (in italian), University of Salerno, May, 1991.
16. DOUGILL J.W., RIDA M.A.M., Further Consideration of Progressively Fracturing Solids, Journal of Engineering Mechanics, ASCE, 1980.
17. FAJFAR P., FISCHINGER M., A Seismic Procedure Including Energy Concept, Proceedings of IX ECEE, Moscow, September, Vol. 2, 312-321, 1990.
18. FAJFAR P., VIDIC T., FISCHINGER M., Seismic Demand in Medium- and Long-Period Structures, Earthquake Engineering and Structural Dynamics, Vol.18, 1133-1144, 1989.
19. FAJFAR P., VIDIC T., FISCHINGER M., A Measure of Earthquake Motion Capacity to Damage Medium-Period Structures, Soil Dynamics and Earthquake Engineering, Vol.9, No.5, 236-242, 1990.
20. HATAMOTO H., CHUNG Y.S., SHINOZUKA M., Seismic Capacity Enhancement of R/C Frames by means of Damage Control Design, Proceedings of Fourth U.S. National Conference on Earthquake Engineering, May 20-24, Palm Springs, Vol. 2, 279-288, 1990.
21. IEMURA H., Earthquake Failure Criteria of Deteriorating Hysteretic Structures, Proceedings of VII WCEE, Istambul, Vol. 5, 81-88, 1980.
22. KRAJCINOVIC D., LEMAITRE J. (Editors), Continuum Damage Mechanics. Theory and Application, CISM, Courses and Lectures, No. 295, Springer Verlag, 1988.
23. KRAWINKLER H., ZOHREI M., Cumulative Damage in Steel Structures Subjected to Earthquake Ground Motion, Computers & Structures, Vol. 16, No. 1-4, 531-541, 1983.

24. KRAWINKLER H., Performance Assessment of Steel Components, Earthquake Spectra, Vol. 3, No. 1, 27-41, 1987.

25. KUNNATH S.K., REINHORN A.M., PARK Y.J., Analytical Modeling of Inelastic Seismic Response of R/C Structures, Journal of Structural Engineering, ASCE, Vol. 116, No. 4, 996-1017, 1990.

26. LASHKARI-IRVANI B., KRAWINKLER H., Damage Parameters for Bilinear Single Degree of Freedom Systems, Vol. 3, 5-16, ECEE, Athen, 1982.

27. MAHIN S., BERTERO V., An Evaluation of Inelastic Seismic Design Spectra, Journal of the Structural Division, ASCE, Vol. 107, No. ST9, 1777-1795, 1981.

28. MEYER I.F., KRATZING W.B., STANGERBERG F., MESKOURIS K., Damage Prediction in Reinforced Concrete Frames Under Seismic Actions, European Earthquake Engineering, No. 3, 9-15, 1988.

29. MURAKAMI M., PENZIEN J., Nonlinear Response Spectra for Probabilistic Seismic Design and Damage Assesment of Reinforced Concrete Structures, Report No. EERC 75-38, November, 1975.

30. PARK Y.J., Seismic Damage Analysis and Damage-Limiting Design of R/C Structures, Ph.D. Thesis, Dept. of Civil Engineering, University of Illinois, Urbana, IL, 1984.

31. PARK Y.J., ANG A.H-S., Mechanistic Seismic Damage Model for Reinforced Concrete, Journal of Structural Engineering, ASCE, Vol. 111, No. 4, 722-739, 1985.

32. PARK Y.J., ANG A. H-S., WEN Y.K., Seismic Damage Analysis of Reinforced Concrete Buildings, Journal of Structural Engineering, ASCE, Vol. 111, No. 4, 740-757, 1985.

33. PARK Y.J., ANG A. H-S., WEN Y.K., Damage-Limiting Aseismic Design of Buildings, Earthquake Spectra, Vol. 3, No. 1, 1-26, 1987.

34. POWELL G.H., ALLAHABADI R. Seismic Damage Prediction by Deterministic Methods: Concepts and Procedures, Earthquake Engineering and Structural Dynamics, Vol.16, 719-734, 1988.

35. REINHORN A.M., MANDER J.B., BRACCI J., KUNNATH S.K., A Post-Earthquake Damage Evaluation Strategy for R/C Buildings, Proceedings of Fourth U.S. National Conference on Earthquake Engineering, May 20-24, Palm Springs, Vol. 2, 1047-1056, 1990.

36. RESENDE L., MARTIN J.B., A Progressive Damage Continuum Model for Granular Materials, Journal of Computer Methods in Applied Mechanics, Vol. 42, 1984.

37. SAIIDI M., SOZEN M.A., Simple Nonlinear Seismic Analysis of R/C Structures, Journal of the Structural Division, ASCE, No. ST5, 937-952, 1981.

38. STEPHENS J.E., Structural Damage Assesment Using Response Measurements, Ph.D. Thesis, Purdue University, 1984.

39. STEPHENS J.E., YAO J.T.P., Damage Assesment Using Response Measurements, Journal of Structural Engineering, ASCE, Vol. 113, No. 4, 787-801, 1987.

40. TEMBULKAR J.M., NAU J.M., Inelastic Modeling and Seismic Energy Dissipation, Journal of Structural Engineering, ASCE, Vol. 113, No. 6, 1373-1377, 1987.

41. THYAGARAJAN R.S., IWAN W.D., Performance Characteristics of a Widely Used Hysteretic Model in Structural Dynamics, Proceedings of Fourth U.S. National Conference on Earthquake Engineering, May 20-24, Palm Springs, Vol. 2, 177-186, 1990.

42. UANG C.M., BERTERO V.V., Evaluation of Seismic Energy in Structures, Earthquake Engineering and Structural Dynamics, Vol.19, 77-90, 1990.

43. WEN Y.-K., Method for Random Vibration of Hysteretic Systems, Journal of Engineering Mechanics, ASCE, Vol. 102, No. EM2, 249-263, 1976.

44. WEN Y.-K., Equivalent Linearization for Hysteretic Systems Under Random Excitation, Journal of Applied Mechanics, ASME, Vol. 47, 150-154, 1980.

45. YAMADA M. et al., Fracture Ductility of Structural Elements and of Structures, paper 6-3-3, Vol. IV, IX WCEE, Tokyo-Kyoto, August 2-9, 1988.

46. YAMADA M., Low Cycle Fatigue Fracture Limits of Structural Materials and Structural Elements, RILEM Workshop on "Needs in Testing Materials", Naples, May 1990.

47. ZAHRAH T., HALL J., Earthquake Energy Absorption in SDOF Structures, Journal of Structural Engineering, ASCE, Vol. 110, No. 8, 1757-1772, 1984.

24. KRAWINKLER H., Performance Assessment of Steel Components, Earthquake Spectra, Vol. 3, No. 1, 27-41, 1987.

25. KUNNATH S.K., REINHORN A.M., PARK Y.J., Analytical Modeling of Inelastic Seismic Response of RC Structures, Journal of Structural Engineering, ASCE, Vol. 116, No. 4, 996-1017, 1990.

26. LASHKARI B., KRAWINKLER H., Damage Parameters for Bilinear Single Degree of Freedom Systems, Vol. 5, Stanford University, 1989.

27. MAHIN S., BERTERO V.V., Problems in Establishing and Predicting Ductility in Seismic Design, ASCE, Vol. 107, No. ST9, 1981.

28. MANDER J.B., PRIESTLEY M.J.N., STANDARD R.J.N., MEMBRANE Approach for the Analysis of Reinforced Concrete Frames, Journal of Structural Engineering, Vol. ?, ?, 1988.

29. MARGANTA J., PINCHEIRA J., Recommendations for Selecting the Probabilistic Response, Design and Damage Assessment of Reinforced Concrete Structures, Report No., ?, 1996.

30. MARK V.J., Seismic Analysis and Damage-Limiting Design of RC Structures, Ph.D. Thesis, Dept. of Civil Engineering, University of Illinois, Urbana, IL, 1984.

31. PARK Y.J., ANG A.H.S., Mechanistic Seismic Damage Model for Reinforced Concrete, Journal of Structural Engineering, ASCE, Vol. 111, No. 4, 722-739, 1985.

32. PARK Y.J., ANG A.H.S., WEN Y.K., Seismic Damage Analysis of Reinforced Concrete Buildings, Journal of Structural Engineering, ASCE, Vol. 111, No. 4, 740-757, 1985.

33. PARK Y.J., ANG A.H.S., WEN Y.K., Damage-Limiting Aseismic Design of Buildings, Earthquake Spectra, Vol. 3, No. 1, 1-26, 1987.

34. POWELL G.H., ALLAHABADI R., Seismic Damage Prediction by Deterministic Methods, Concepts and Procedures, Earthquake Engineering and Structural Dynamics, Vol. 16, 719-734, 1988.

35. REINHORN A.M., MANDER J.B., BRACCI J.M., KUNNATH S.K., A RC Experimental Verification and Analytical Studies for the Seismic Response of Steel, U.S.-Japan Conference on Damage Detection, Proceedings, Lake Sunapee, Vol. 2, 1987-1990, ?.

36. REINHORN A.M., MARTIN P., A Theory of Hysteresis Behavior Model for Granular Materials, Journal of the Engineering Mechanics Division, ASCE, Vol. ?, ?, 1984.

37. SAIIDI M., SOZEN M.A., Simple Nonlinear Seismic Analysis of R.C. Structures, Journal of the Structural Division, ASCE, Vol. 107, No. ST5, 937-953, 1981.

38. STEPHENS J.E., Structural Damage Assessment Using Response Measurements, Ph.D. Thesis, Purdue University, 1987.

39. STEPHENS J.E., YAO J.T.P., Damage Assessment Using Response Measurements, Journal of Structural Engineering, ASCE, Vol. 113, No. 4, 787-801, 1987.

40. TOUSSI S., YAO J.T.P., Hysteresis Identification of Existing Structures, Journal of the Engineering Mechanics Division, ASCE, Vol. 109, No. 5, 1971, 1983.

41. WARAN A.M., SOZEN M.A., Failures, Close Encounters and Lessons Learned in Weak Story Mechanisms, Structural Failure and Remedial Measures, Proceedings, ?, 1989.

42. WEN Y.K., Response and Damage of Structures Subjected to Random Excitation, 9th World Conference on Earthquake Engineering, Proceedings, May 2-9, Tokyo, Kyoto, Japan, Vol. 8, 1988.

43. WILLIAMS M.S., SEXSMITH R.G., Seismic Damage Indices for Concrete Structures: A State-of-the-Art Review, Earthquake Spectra, Vol. 11, No. 2, 319-349, 1995.

44. WEN Y.K., Method for Random Vibration of Hysteretic Systems, Journal of the Engineering Mechanics Division, ASCE, Vol. 102, No. EM2, 249-263, 1976.

45. WEN Y.K., Equivalent Linearization for Hysteretic Systems Under Random Excitation, Journal of Applied Mechanics, ASME, Vol. 47, 150-154, 1980.

46. YAMADA M., et al., Fracture Mechanics and Damage Mechanics for Structures, Proceedings of ?, ?, No. ?, 1988.

47. YAMADA M., et al., Fracture Mechanics and Damage Mechanics for Structural and Structural Materials, 9th World Conference on Earthquake Engineering, Proceedings, Tokyo, May 1988.

48. ZERVA A.L., ANG A.H.S., Earthquake Damage Assessment Incorporating Uncertainty, Journal of Structural Engineering, ASCE, Vol. 110, No. 8, 1757-1772, 1984.

SEISMIC ENERGY DEMANDS ON REINFORCED CONCRETE DUCTILE MOMENT-RESISTING FRAME BUILDINGS

T.J. Zhu, W.K. Tso and A.C. Heidebrecht
Department of Civil Engineering, McMaster University
Hamilton, Ontario, Canada L8S 4L7

ABSTRACT

This paper examines the seismic energy demands on reinforced concrete ductile moment-resisting frame buildings and the feasibility of using equivalent single degree of freedom systems to estimate these energy demands. Three buildings having 4, 10 and 18 storeys were designed to the current Canadian seismic provisions; three groups of earthquake records having different frequency contents were selected as input ground motions. The input, hysteretic, and damping energies were evaluated for the three buildings and their equivalent single degree of freedom systems when they were subjected to the three groups of ground motions.

INTRODUCTION

The seismic energy imparted into a building structure during strong earthquake ground motion is dissipated in part by damping and in part by yielding of the structural components. Proper seismic design implies that the energy dissipation capacity of the building structure should exceed the energy absorption demand imposed by the ground motion. Since the early work by Housner [1], the seismic energy concept has gained extensive attention [2-8]. Most of these previous studies were concerned with the evaluation of seismic energy demands on single degree of freedom (SDOF) systems. The objectives of this study are (1) to extend the seismic energy concept to multi degree of freedom (MDOF) systems, specifically multistorey reinforced concrete ductile moment-resisting frame (DMRF) buildings, and (2) to study the feasibility of using equivalent SDOF systems to estimate seismic energy demands on these multistorey buildings.

STRUCTURAL MODELS

Three reinforced concrete DMRF buildings having 4, 10, and 18 storeys were considered. These three buildings are denoted as 4S, 10S, and 18S, and they have the same floor plan as shown in Fig. 1. The effect of seismic action was considered in the N-S direction. The elevations of the interior frames in this direction are shown in Fig. 1. The beam and column sizes are also depicted. The three buildings were designed for combined gravity and seismic effects in accord with the 1990 edition of the National Building Code of Canada [9], and their structural members were proportioned and detailed according to the 1984 edition of the Canadian Concrete Code [10]. A detailed account of the design procedure and the final design results of the three frames is given by Zhu [11]. The fundamental period of the 4S, 10S, and 18S frames is 0.47, 1.20, and 2.01 sec, respectively.

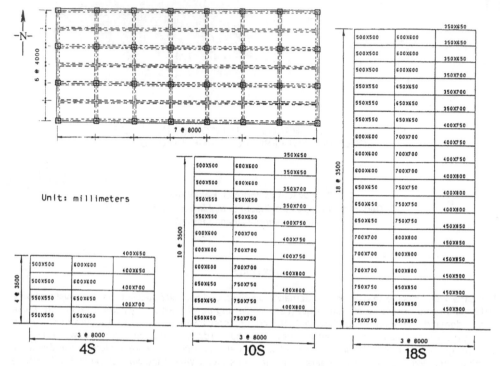

Unit: millimeters

Figure 1. Floor plan and frame elevation.

GROUND MOTION DATA

A total of 45 strong motion records recorded on rock or stiff soil sites were selected from the McMaster University Seismological Executive (MUSE) Database. The 45 records were obtained from 23 different events with magnitude ranging from 5.25 to 8.1, and they were recorded at epicentral distances ranging from 4 to 379 km. The 45 records were subdivided into three groups according to their peak acceleration-to-velocity (A/V) ratios, with 15 records in each group. The records having low, intermediate, and high A/V ratios are taken as representative ground motions in seismic regions in Canada designated as $Z_a < Z_v$, $Z_a = Z_v$, and $Z_a > Z_v$, respectively. Fig. 2 shows the distribution of magnitude and epicentral distance for the three A/V groups of records. It can be seen that the records with high A/V ratios were obtained in the vicinity of small or moderate earthquakes whereas those having low or intermediate A/V ratios were recorded at large distances from large or moderate earthquakes. A detailed description and analysis of the earthquake data is given by Zhu [11].

DYNAMIC ANALYSIS PROCEDURE

The general-purpose computer program DRAIN-2D [12] was used to perform dynamic analysis for the frames. The dual-component element was used to model the beams and columns. The effect of axial force on yield moment was considered for each column by a yield moment-axial force interaction curve. A simplified solution [12] was used to account for the P-delta effect for each column. The damping matrix was expressed as a linear combination of the mass and stiffness matrices, and the combination coefficients were selected to give 5% of critical damping in the first two vibrational modes.

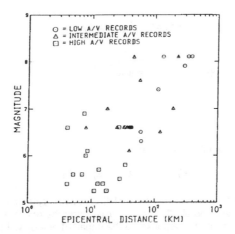

Figure 2. Distribution of magnitude and epicentral distance for three A/V groups of records.

SEISMIC ENERGY EQUATION

The equation of motion for a MDOF system is given by

$$[M]\{\ddot{u}\} + [C]\{\dot{u}\} + \{R(\{u\})\} = -[M]\{r\}\ddot{u}_g \qquad (1)$$

in which [M] = the mass matrix; [C] = the damping matrix; $\{R(\{u\})\}$ = the restoring force vector; $\{\ddot{u}\}$, $\{\dot{u}\}$, and $\{u\}$ = the relative acceleration, velocity, and displacement vector; $\ddot{u}_g$ = the ground acceleration; and $\{r\}$ = a vector relating the ground acceleration to the structural degrees of freedom. Pre-multiplying Eq.(1) by the transposed incremental relative displacement vector, $\{du\}^T = \{\dot{u}\}^T dt$, leads to

$$\{\dot{u}\}^T[M]\{\ddot{u}\} dt + \{\dot{u}\}^T[C]\{\dot{u}\} dt + \{\dot{u}\}^T\{R(\{u\})\} dt = -\{\dot{u}\}^T[M]\{r\}\ddot{u}_g dt \qquad (2)$$

Integration of Eq.(2) with respect to time, t, yields

$$\int_0^t \{\dot{u}\}^T[M]\{\ddot{u}\} dt + \int_0^t \{\dot{u}\}^T[C]\{\dot{u}\} dt + \int_0^t \{\dot{u}\}^T\{R(\{u\})\} dt = -\int_0^t \{\dot{u}\}^T[M]\{r\}\ddot{u}_g dt \qquad (3)$$

The first term of Eq.(3) is the kinetic energy, E_K^*:

$$E_K^* = \int_0^t \{\dot{u}\}^T[M]\{\ddot{u}\} dt \qquad (4)$$

The second term in Eq.(3) is the damping energy, E_D^*:

$$E_D^* = \int_0^t \{\dot{u}\}^T[C]\{\dot{u}\} dt \qquad (5)$$

The third term of Eq.(3) is the sum of the irrecoverable hysteretic energy, E_H^*, and the recoverable elastic strain energy, E_S^*:

$$E_H^* + E_S^* = \int\limits_0^t \{\dot{u}\}^T \{R(\{u\})\} \, dt \tag{6}$$

The right-hand-side term in Eq.(3) is the input energy, E_I^*:

$$E_I^* = -\int\limits_0^t \{\dot{u}\}^T [M] \{r\} \ddot{u}_g \, dt \tag{7}$$

The energy equation (Eq.(3)) can then be written as

$$E_I^* = E_K^* + E_D^* + E_H^* + E_S^* \tag{8}$$

Eq.(8) indicates that during the seismic response of the system, part of the input energy is stored temporarily in the form of kinetic and elastic strain energies, and the rest is dissipated by damping and inelastic deformation. At the end of the response, all the imparted energy will be dissipated by damping and hysteretic action. Eq.(8) can be normalized to the total mass of the system. The energy quantities per unit mass are denoted as E_I, E_K, E_D, E_H, and E_S, and their numerical computations have been incorporated into the DRAIN-2D program [11].

SEISMIC ENERGY DEMANDS ON MDOF MODELS

Fig. 3 shows the energy time-history response of the 10S frame subjected to the site 1 record of the 1985 Nahanni earthquake in Canada. Table 1 presents the mean values of the input, damping, and hysteretic energies for the three frames subjected to the three A/V groups of ground motions. All the input ground motions were scaled to a common peak velocity of 0.4 m/s which is the design zonal velocity for the frames. The damping and hysteretic energies are expressed as a percentage of the total dissipated energy.

The values of the input and hysteretic energies for the three frames generally reflect their overall seismic performance. For the 4S frame, the beam ductility demands for the high and intermediate A/V groups of ground motions are about 70% higher than those for the low A/V group [11]. It can be seen in Table 1 that the input energy and also the hysteretic energy for the high and intermediate A/V groups of records is about twice that for the low A/V group for this frame. On the other hand, the beam ductility demands for the low and intermediate A/V groups of ground motions are about 100% higher than those for the high A/V group for the 18S frame [11]. Table 1 shows that for the 18S frame, the input energy for the low and intermediate A/V groups of records is almost twice that for the high A/V group, and the hysteretic energy for the low and intermediate A/V groups of records is also much higher than that for the high A/V group.

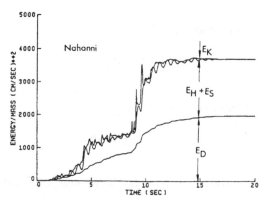

Figure 3. Energy time-history response of 10S frame subjected to a 1985 Nahanni earthquake record.

TABLE 1
Mean values of input energy, damping energy, and hysteretic energy

Energy	Frame	Low A/V	Intermediate A/V	High A/V
E_I $(cm/s)^2$	4S	3144	7486	6231
	10S	5875	7693	4329
	18S	4816	5649	2827
E_D (%)	4S	53	43	42
	10S	53	51	56
	18S	54	60	72
E_H (%)	4S	47	57	58
	10S	47	49	44
	18S	46	40	28

Although the correlation between energy demand and ductility demand is good for both stiff (4S) and flexible (18S) frames, the use of a single energy quantity (input or hysteretic energy) cannot provide information on the distribution of inelastic deformations. For example, the distribution of inelastic deformations over the height of the 10S and 18S frames is significantly different for the three A/V groups of ground motions [11]. For the low A/V group of ground motions which have lower frequency content, inelastic deformations are concentrated in the lower storeys. For the high A/V group of ground motions which have higher frequency content, the higher modal effect becomes significant, and inelastic deformations in the upper storeys are higher than those in the lower storeys.

Previous studies [2,4,8] based on SDOF systems have indicated that the input energy is not sensitive to the level of inelastic response (the level of displacement ductility), and the input energy for elastic response is similar to that for inelastic response. This is particularly true for systems having relatively long periods. In this study, the input energy for the elastic response of the three frames was also computed. The elastic input energy was computed by setting $\{R(\{u\})\} = [K]\{u\}$ in Eq.(3), where $[K]$ = the initial elastic stiffness matrix of a frame. Fig. 4 shows a comparison between the input energy for inelastic response, $(E_I)_{inelastic}$, and the input energy for elastic response, $(E_I)_{elastic}$, for the 45 individual earthquake records. Table 2 presents the ratios between the mean values of $(E_I)_{inelastic}$ and $(E_I)_{elastic}$ for the three A/V groups of ground motions. It can be seen that the input energy for elastic response is similar to that for inelastic response. This is particularly true for the 18S frame. Therefore, one can speculate that the similarity between elastic and inelastic input energy for SDOF systems also holds for MDOF systems representative of regular DMRF buildings.

TABLE 2
Mean ratios of inelastic input energy to elastic input energy

Frame	Low A/V	Intermediate A/V	High A/V
4S	1.13	1.01	0.82
10S	0.85	0.93	0.88
18S	0.94	1.02	0.96

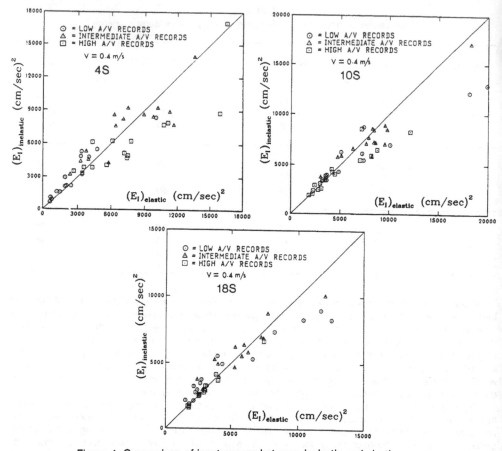

Figure 4. Comparison of input energy between inelastic and elastic responses.

To examine the sensitivity of the inelastic response of the frames to an increase in the peak ground velocity level, the peak velocity for scaling the input ground motions was increased from the design level of 0.4 m/s to a level of 0.6 m/s. Fig. 5 compares the input energy for the peak ground velocity level of 0.6 m/s, $(E_I)_{0.6m/s}$, with the input energy for the peak ground velocity level of 0.4 m/s, $(E_I)_{0.4m/s}$, for the 45 individual records. Table 3 presents the ratios of the mean values between $(E_I)_{0.6m/s}$ and $(E_I)_{0.4m/s}$ for the three A/V groups of ground motions. It can be seen in Eq.(7) that if the response of a frame remains in the elastic range, the scaling of the input ground motion by a factor of λ will result in a corresponding scaling factor of λ^2 for the input energy. Fig. 5 and Table 3 show that the ratios of $(E_I)_{0.6m/s}$ to $(E_I)_{0.4m/s}$ are very close to the scaling factor of $(0.6/0.4)^2 = 2.25$. Therefore, it appears that the scaling factor for elastic input energy may be used for inelastic input energy, as noted by Jennings [2] in the case of SDOF systems. Table 3 indicates that the increase in the hysteretic energy ratio is larger than λ^2 whereas the increase in the damping energy ratio is less than λ^2. In other words, most of the increased input energy resulting from the increased level of the input ground motions had to be dissipated through hysteretic action. This is consistent with the observed substantial increase in the beam and column ductility demands in the lower storeys. The 50% increase in the peak ground velocity level has led to almost 100% increase in the ductility demands on the beams and columns of the lower storeys [11].

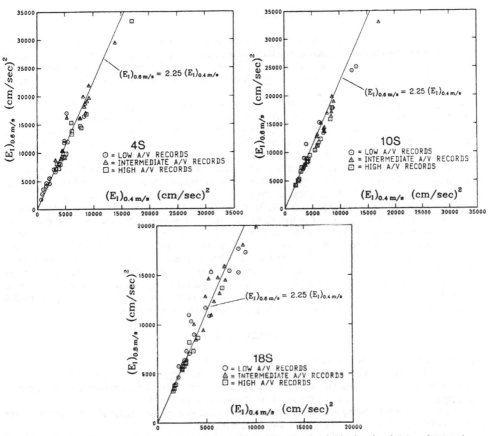

Figure 5. Comparison of input energy between peak ground velocity levels of 0.4 and 0.6 m/s.

TABLE 3
Mean ratios of energy indices between peak ground velocity levels of 0.6 and 0.4 m/s

Energy	Frame	Low A/V	Intermediate A/V	High A/V
E_I	4S	2.46	2.23	2.02
	10S	2.11	2.06	2.03
	18S	2.29	2.22	2.14
E_D	4S	1.75	1.77	1.71
	10S	1.60	1.63	1.65
	18S	1.76	1.74	1.77
E_H	4S	3.27	2.59	2.22
	10S	2.67	2.50	2.45
	18S	2.91	2.94	3.01

SEISMIC ENERGY DEMANDS ON EQUIVALENT SDOF SYSTEMS

Most previous studies [2-8] on seismic energy demands were for SDOF systems. The applicability of the results of these studies to actual building structures depends on the extent to which the energy demands for multistorey buildings will correlate with those for SDOF systems. The second objective of this study is therefore to examine the possibility of using equivalent SDOF systems to estimate seismic energy demands for regular DMRF buildings.

Saiidi [13] suggested a simplified analysis procedure to estimate inelastic deformation for regular reinforced concrete building structures. In this procedure, an actual multistorey building is transformed into an equivalent SDOF system. The force-deformation behaviour of this equivalent SDOF systems is obtained from an inelastic static analysis of the actual building subjected to monotonically increased lateral loading. In this study, the 4S, 10S, and 18S frames were converted into their equivalent SDOF systems based on Saiidi's approach. A detailed description of this conversion for the three frames can be found in Zhu [11]. The various energy terms were then computed for the equivalent SDOF systems, and they were compared with those obtained from the MDOF models of the three frames.

Fig. 6 compares the input energy for the MDOF models of the three frames, $(E_i)_{MDOF}$, with that of their equivalent SDOF systems, $(E_i)_{SDOF}$. For the 4S frame, the input energy estimated from its equivalent SDOF system is very close to that obtained from its MDOF model for all 45 earthquake records. However, this correlation between $(E_i)_{MDOF}$ and $(E_i)_{SDOF}$ deteriorates for the 18S frame. This is particularly true when the 18S frame is subjected to the high and intermediate A/V ground motions which have relatively high frequency contents. The same observation can be made regarding to hysteretic energy comparisons as shown in Fig.7. For the 18S frame, the contribution of higher modal responses becomes significant [11]. The use of an equivalent SDOF system cannot account for the higher modal contributions, and consequently it leads to poor estimates of the energy demands.

CONCLUSIONS

Based on this study, the following conclusions can be drawn:

(1) The values of the input and hysteretic energy demands for regular ductile moment-resisting frame (DMRF) buildings generally reflect their overall seismic performance. But they cannot provide information regarding the distribution of inelastic deformations throughout the buildings.

(2) The elastic input energy is similar to the inelastic input energy for regular DMRF buildings. This is particularly true for buildings having relatively long periods.

(3) For the elastic response of a building, the scaling of the input ground motion by a factor of λ will lead to a corresponding scaling factor of λ^2 for the input energy. This scaling factor for the elastic input energy seems also applicable to the inelastic input energy for regular DMRF buildings.

(4) Equivalent SDOF systems may be used to estimate the input and hysteretic energy demands on low-rise DMRF buildings. For high-rise DMRF buildings, the contributions of higher modal responses become significant, and the use of equivalent SDOF systems may underestimate the energy demands on the buildings. This is particularly true for input ground motions having relatively high frequency content which will accentuate the higher modal contributions.

ACKNOWLEDGEMENT

The writers wish to acknowledge the support from the Natural Sciences and Engineering Research Council of Canada for the work presented herein.

103

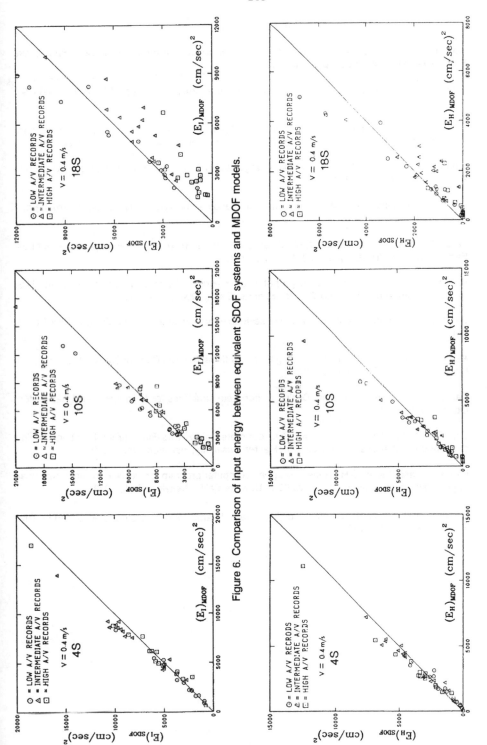

Figure 6. Comparison of input energy between equivalent SDOF systems and MDOF models.

Figure 7. Comparison of hysteretic energy between equivalent SDOF systems and MDOF models.

REFERENCES

1. Housner, G.W., Limit design of structures to resist earthquakes. Proc. 1st World Conf. Earthquake Eng., Berkeley, Calif., 1956, pp. 5-1 to 5-13.

2. Jennings, P. C., Earthquake response of a yielding structure. J. Eng. Mech. Div., ASCE, 1965, 91, 41-68.

3. McKevitt, W.E., Anderson, D.L., Nathan, N.D. and Cherry S., Towards a simple energy method for seismic design of structures. Proc. 2nd U.S. Nat. Conf. Earthquake Eng., 1979, pp. 383-392.

4. Zahrah, T. F. and Hall, W. J., Earthquake energy absorption in SDOF structures. J. Struct. Eng., ASCE, 1984, 110, 1757-1772.

5. Tembulkar, J.M. and Nau, J.M., Inelastic modelling and seismic energy dissipation. J. Struct. Eng., ASCE, 1987, 113, 1373-1377.

6. Fajfar, P., Vidic, T. and Fischinger, M., Seismic demand in medium- and long-period structures. Earthquake Eng. Struct. Dyn., 1989, 18, 1133-1144.

7. Krawinkler, H. and Nassar, A., Damage potential of earthquake ground motions. Proc. 4th U.S. Nat. Conf. Earthquake Eng., Vol.2, Palm Springs, Calif., 1990, pp. 945-954.

8. Uang, C.M. and Bertero, V.V., Evaluation of seismic energy in structures. Earthquake Eng. Struct. Dyn., 1990, 19, 77-90.

9. Associate Committee on National Building Code, National Building Code of Canada 1990. National Research Council, Ottawa, Ont., 1990.

10. Canadian Standards Association, Design of concrete structures for buildings, CAN3-A23.3-M84. Ottawa, Ont., 1984.

11. Zhu, T.J., Inelastic response of reinforced concrete frames to seismic ground motions having different characteristics. Ph.D. thesis, McMaster University, Hamilton, Ont., 1989.

12. Kanaan, A. and Powell, G.H., General purpose computer program for inelastic dynamic response of plane structures. Report No. EERC 73-6, Univ. of Calif., Berkeley, Calif., 1973.

13. Saiidi, M. and Sozen, M. A., Simple nonlinear seismic analysis of R/C structures. J. Struct. Div., ASCE, 1981, 107, 937-952.

RESPONSE OF REINFORCED CONCRETE MOMENT FRAMES
TO STRONG EARTHQUAKE GROUND MOTIONS

HIROSHI AKIYAMA and MAKOTO TAKAHASHI
Department of Architecture
Faculty of Engineering, University of Tokyo,
Hongo, Bunkyo-ku, Tokyo, Japan

ABSTRACT

Using restoring-force characteristics peculiar to the reinforced concrete rigid frame, inelastic responses of multi-story rigid frames to strong seismic ground motions were analysed. The relation between the energy input and the damage distribution was made clear, and it was found that there exist no essential differences between steel structures and reinforced concrete structures in the manner of resisting earthquakes.

INTRODUCTION

The design method based on the energy concept in which the loading effect of earthquakes on structures is considered to be the energy input and the structural resistance is considered to be the energy absorption capacity has been proved to be effectively applied to steel structures . On the other hand, researches on the energy response of reinforced concrete structures have been developed considerably, but they remain still insufficient to construct design criteria[2),3)].
The difference of structural response between steel structures and reinforced structures is rooted from the difference of their restoring force characteristics.
In steel structures, the fundamental types of restoring force characteristics are the elastic-perfectly plastic type and the slip type, and the actual restoring force chracteristics takes very complicated shapes as is found in the mixed structure of moment frames and braced frames. Nevertheless, even for such a complicated structure, the design method based on the energy concept can be applied.
Since the restoring force characteristics of reinforced concrete structures are not quitely different from those of steel structures, the methodology which is applicable to the steel structures can be also applied to the reinforced concrete structures.
In this paper, the prototype of restoring force characteristics of reinforced concrete moment frames is introduced and to what extent the

design criteria must be modified is made clear. In previous researches of reinforced concrete structures, especially the damage distribution in multi-story buildings and the damage concentration have been scarcely discussed. The major emphasis is laid on these items.

The earthquake record used in the analysis is Hachinohe record of the Tokachi-oki earthquake (1968). The maximum ground acceleration is 183gal. The vibrational model is of the shear type which corresponds to the weak column type moment frame.

ANALYTICAL MODEL

Vibrational System

In Fig.1 the vibrational model is shown. It is assumed that only columns deform and masses are concentrated on the rigid beams, the axial deformation of columns are neglected. Masses are evenly distributed. The yield story displacement in each story is assumed to be equal.

Restoring Force Characteristics

The relationship between the story shear force Q_i and the story displacement δ_i is shown in Fig.2. This type of restoring force characteristics is typical of the reinforced concrete beam-columns and has been referred by many researchers[2],[3]. The $Q_i - \delta_i$ relationship under the monotonic loading forms a skeleton curve in the restoring-force characteristics. Under the monotonic loading, the frame remains elastic for the load below the bending crack initiation load Q_{ci} with the spring constant in the elastic range K_{Ei}. The yield strength reached by the yielding of reinforcing bars is denoted by Q_{yi}. The deformation which corresponds to Q_{yi} is defined as the yield story displacement and denoted by δ_{yi}. The secant slope of the segment which connects the point of origin and the yield point (Q_{yi}, δ_{yi}) is denoted by K_i. As δ_{yi} is assumed constant, K_i becomes proportional to Q_{yi}. The slope of the skeleton curve beyond the yield point is assumed flat.

Q_{yi}/Q_{ci} and δ_{yi}/δ_{ci} are assumed as

$$\left.\begin{array}{l} \dfrac{Q_{yi}}{Q_{ci}} = 3.0 \\[2mm] \dfrac{\delta_{yi}}{\delta_{ci}} = 10.0 \end{array}\right\} \tag{1}$$

Therefore, K_i/K_{Ei} becomes as follows.

$$\dfrac{K_i}{K_{Ei}} = 0.3 \tag{2}$$

The hysteresis rule is demonstrated in Fig.2, following the progress of deformation from the point 1 to 7.

PRINCIPAL RESPONSES

Total Energy Input

The total energy input to the one-mass system equipped with the restoring force characteristics shown in Fig.2 is shown Fig.3 where the total energy

input is expressed by the equivalent velocity
($V_\varepsilon = \sqrt{2E/M}$ E: the total energy input, M: mass).
E is defined as

$$E = -\int_0^{t_0} m\ddot{y}_0 \dot{y}\, dt \qquad (3)$$

where y_0 : ground motion
      y : the relative displacement of the mass
      t_0 : the duration of ground motion

The abscissa indicates the natural period which is obtained by using the
secant stiffness K_i in Fig.2 ($T = 2\pi\sqrt{M/K_i}$).
As   an index of the plastification of system the average plastic
deformation ratio $\bar{\mu}$ is used. $\bar{\mu}$ is defined as

$$\bar{\mu} = \frac{\delta_{max}^+ + \delta_{max}^-}{2\delta_Y} - 1 \qquad (4)$$

where $\delta_{max}^+, \delta_{max}^-$: the maximum story displacement in the positive and
negative direction respectively.

In Fig.3, V_ε – values for $\bar{\mu}$ = 1,2,5, are shown. The solid line in the
figure is the V_ε – T relationship for the elastic system with h = 0.1, and
the broken lines are line segments which envelope it.   This broken lines
have been proposed by the author as the energy spectrum for design use[1].
The slope of the line segment which starts from the point of origin is
increased to be 1.2 times as large as that of the original slope on the
reason that the energy input into the inelastic system with shorter
natural periods tends to be greater than that of the elastic system. It is
clearly observed that the energy input into the system with a period
shorter than 0.5sec tends to increase beyond the design spectrum as
$\bar{\mu}$ increases, while the design spectrum almost covers the energy response
of the system with a period longer than 0.5sec. To meet the increase of
energy input in the range of shorter natural period, the design spectrum
must be increased in such a manner as shown in Fig.3, and the modified
value can be written by

$$V_\varepsilon = 1.34\, V_{E0} \qquad (5)$$

where V_{E0} : the value of design spectrum.

ENERGY ABSORPTION CAPACITY

The relation between the average inelastic deformation ratio $\bar{\mu}$ and the
total energy input E is investigated.   The $\bar{\mu} - E$ relationship can be
formally expressed as

$$E = 2a_s\bar{\mu}Q_Y\delta_Y \qquad (6)$$

Using the values of E and $\bar{\mu}$ obtained by the response analysis for one
mass systems, a_s is calculated from Eq(6) and is shown in Fig.4.
The value of a_s tends to decrease as the natural period increases.
To estimate the lower bound rule of a_s, a deformation history shown in
Fig.5 is assumed.   This history does not include the initial nonlinear
path which leads to the yield point, and experiences the same amount of

inelastic deformations in the positive and negative directions.
The accumulated strain energy at the point a must be equal to the total energy input, and the following equation is obtained

$$E = Q_r \delta_r \left\{ \frac{1}{2} + 2\bar{\mu} \left(1 + \frac{3}{4} \right) \right\}$$

(7)

Therefore, the value of a_s which corresponds to the deformation history shown in Fig.5 is obtained as

$$a_s = \frac{1}{4\bar{\mu}} + 1.75$$

(8)

a_s calculated from Eq(8) is shown in Fig 4.
It is seen that Eq(8) gives a lower bound value of a_s in the range of period shorter than 4.0sec.
Referring to Eq(7), the cumulative inelastic strain energy W_p which corresponds to the deformation history shown in Fig.5 becomes

$$W_p = 3.5\bar{\mu} Q_r \delta_r$$

(9)

The value of a_s of the system with a period shorter than 0.5sec is far greater than the value given by Eq(8). On the other hand, the system with a shorter natural period must receive a greater input energy compared to the design spectrum as shown in Fig.3.
Referring to Eq(5), the increased input energy is 1.8 times as large as that specified by the design spectrum.
To meet the increase of input energy, the system must be equipped with the energy absorption capacity 1.8 times as large as that of the system with a longer natural period.
The required level of the energy absorption capacity can be expressed in terms of W_p as

$$W_p = 3.5\bar{\mu} Q_r \delta_r \times 1.8$$

(10)

The value of a_s which corresponds to Eq(10) is given by

$$a_s = \frac{1}{4\bar{\mu}} + 3.15$$

(11)

The a_s — value given by Eq(11) is also indicated in Fig.4. a_s — values for shorter periods are still greater than that given by Eq(11).
Thus, it is concluded that the increase of energy input in shorter period range is well compensated by the increase of energy absorption capacity of the system, and the design spectrum and the estimate of energy absorption capacity expressed by Eq(9) are effectively used regardless of the range of natural period.

OPTIMUM SHEAR FORCE COEFFICIENT DISTRIBUTION

The damage of each story of multi-story frames is defined by

$$W_{pi} = \int_0^{\cdot\delta_i} Q_i d\delta_i$$

(12)

where $\cdot\delta_i : \delta_i$ at the end of the duration ground motion.
The cumulative plastic deformation ratio of each story is defined by

$$\eta_i = \frac{W_{pi}}{Q_{yi}\delta_{yi}} \tag{13}$$

For the shear type multi-story frames equipped with the elastic – perfectly plastic type of restoring force characteristics, a yield shear force coefficient distribution which makes η_i constant exists and the distribution is approximately given by

$$\overline{a}_i = f\left(\frac{i-1}{N}\right) \tag{14}$$

$$f(x) = 1 + 1.5927x - 11.852x^2 + 42.583x^3$$
$$- 59.48x^4 + 30.16x^5 \tag{15}$$

where N : the number of story

For the system with the restoring force characteristics shown in Fig.2, it was found that the same distribution applies as the optimum yield shear force coefficient as shown in Fig.6.

Damage Distribution Law
The damage distribution law which can be applied to the multi – story frames with the elastic – perfectly plastic type of restoring force characteristices has been already made clear[1] and is expressed in a form as follows

$$\frac{W_{pk}}{W_p} = \frac{s_k p_k^n}{\sum_{j=1}^{N} s_j p_j^n} \tag{16}$$

where $p_j = (a_j/a_1)/\overline{a}_j$: the deviation of shear force coefficient
distribution from the optimum distribution
a_j : the yield shear force coefficient in jth story
n : the damage concentration index

The value of n for the shear type multi – story frames with the elastic – perfectly plastic restoring force characteristics was already obtained as follows

$$n = -12 \tag{17}$$

The applicability of the value given by Eq(17) to shear type reinforced concrete moment frames is examined.
First, the damage distribution under the optimum yield shear force coefficient is calculated.   Next the yield shear force coefficient in the observed kth story is reduced to $p_d a_k$, leaving the yield shear force coefficient in the stories other than kth story unchanged.   In Fig.7 the damage distribution in the observed kth story W_{pk} is shown in contrast with the damage distribution under the optimum yield shear force coefficient distribution $\cdot W_{pk}$.
Prediction is made by using Eq(16) and is indicated in Fig.7.
From the figure it can be judged that the same value of damage concentration index can be used for steel frames and reinforced concrete frames.

BASIC DESIGN FORMULA

Based on the response characteristics of reinforced concrete structures, the required strength of multi - story frames is derived.
The equilibrium of energy under an earthquake is written as

$$W_e + W_h + W_p = E \tag{18}$$

where W_e : the elastic vibrational energy
W_h : the energy absorption due to damping

As for the energy input attributable to the strain energy $E - W_h$, the following rough estimate can be made[1].

$$E - W_h = \frac{E}{(1 + 3h + 1.2\sqrt{h})^2} \tag{19}$$

Analogously to the expression of strain energy given by Eq(7), the strain energy of the story of a multi - story frame may be expressed by

$$W_i = Q_{ri}\delta_{ri}\left(\frac{1}{2} + 3.5\,\bar{\mu}_i\right) \tag{20}$$

The first term of Eq(20) expresses the elastic strain energy, while the second term expresses the cumulative inelastic strain energy.
Therefore, W_e and W_p in Eq(18) are described as

$$\begin{aligned} W_e &= \sum_{i=1}^{N} W_{ei} = \sum W_{ei} = \sum \frac{Q_{ri}\delta_{ri}}{2} \\ W_p &= \sum_{i=1}^{N} W_{pi} \end{aligned} \Bigg\} \tag{21}$$

When the yield shear force coefficient distribution agrees with the optimum distribution, W_e is written as

$$W_e = \frac{Mg^2 T^2}{4\pi^2} \cdot \frac{a_1^2}{2} \cdot \frac{\sum_{i=1}^{N} s_i}{x_1} \tag{22}$$

where x_1 : $k_1 T^2 / 4\pi^2 M$
T : the fundemental natural period
M : the total mass of a frame

In the case where each story has a constant mass and a constant yield story displacement, $\sum_{i=1}^{N} s_i / x_1$ and x_1 are approximated by

$$\begin{aligned} \frac{\sum_{i=1}^{N} s_i}{x_1} &= 1.07 \\ x_1 &= 0.42 + 0.58 N \end{aligned} \Bigg\} \tag{23}$$

Therefore, Eq(21) can be approximated by

$$W_e = \frac{Mg^2 T^2}{4\pi^2} \cdot \frac{a_1^2}{2} \tag{24}$$

On the other hand, using the expression of $r_1 = W_p / W_{p1}$, W_p is expressed by

$$W_p = W_{p1} r_1 = 3.5 r_1 Q_n \delta_n \bar{\mu}_1 \qquad (25)$$

r_1 can be obtained from Eq(16) by applying suitable value of p_j .
Referring to the practical deviation of strength distribution from the optimum distribution, the following values have been proposed for the practical design purpose

$$p_j = 1 \qquad j \neq 1$$
$$p_1 = \frac{1}{1.185 - 0.0014N} \qquad (26)$$

Substituting Eq(19),(24) and (25) into Eq(18), the required strength for the first story of reinforced concrete multi - story rigid frames is obtained as follows

$$a_1 = \frac{1}{\sqrt{1 + \frac{7 r_1 f_{t_1}}{x_1}}} \cdot \frac{2\pi V_E}{T g} \cdot \frac{1}{1 + 3h + 1.2\sqrt{h}} \qquad (27)$$

where h : the damping constant.

CONCLUSION

The seismic resistance of reinforced concrete rigid frames was discussed on the basis of energy concept.
Conclusions are summarized as follows.

1) The energy spectrum demonstrated in Fig.3 can be generally applied for the estimate of seismic resistance of frames regardless of the structural type.
2) The damage concentration characteristics of reinforced concrete frames are similar to those of steel frames.
3) The design formula for the reinforced concrete rigid frames is expressed by Eq(27).

EFERENCES

1. Akiyama, H.: Earthquake-Resistant Limit-State Design for Buildings, University of Tokyo Press, 1985
2. Minami, T. and Osawa, Y.: Elastic-Plastic Response Spectra for Different Hysteretic Rules, Earthquake Eng. and Structural Dynamics, Vol.16, 555 - 568, 1988
3. Fajfar, P, Vidic, T. and Fischinger, M.: Seismic Demand in Medium-and Long-Period Structures, Earthquake Eng. and Structural Dynamics, Vol.18, 1133 - 1144, 1989

Figure 1. Vibrational model

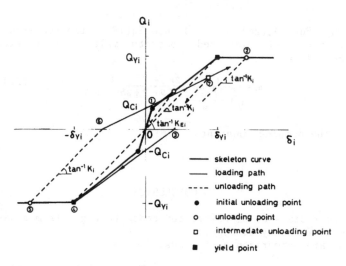

Figure 2. Hysteretic rule

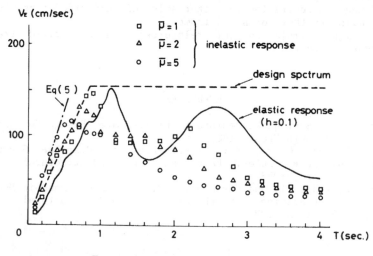

Figure 3. Total energy input

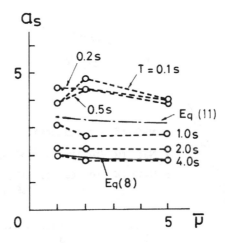

Figure 4. $a_s-\bar{\mu}$ relationship

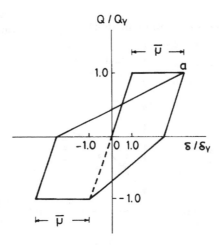

Figure 5. Assumed hysteresis loop

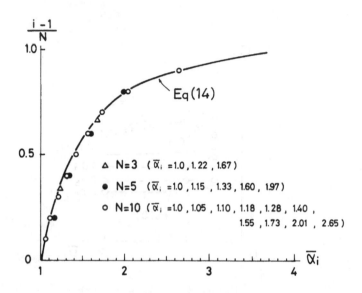

Figure 6. Optimum yield-shear-force coefficient distribution

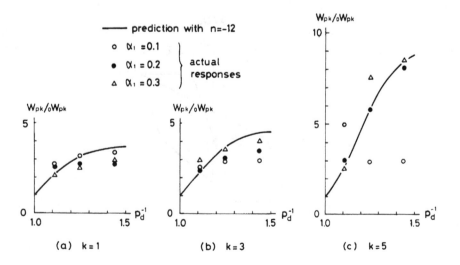

Figure 7. Damage distribution

AN ENERGY-BASED DAMAGE MODEL FOR INELASTIC DYNAMIC ANALYSIS OF REINFORCED CONCRETE FRAMES

CHRISTIAN MEYER

Department of Civil Engineering and Engineering Mechanics
Columbia University, New York, NY 10027

Abstract

An energy-based damage model is introduced, which is suitable for quantifying damage on the material and member level of reinforced concrete structures. It lends itself to incorporation into nonlinear dynamic analysis programs for concrete structures. Experimental low-cycle fatigue data are presented to illustrate the use of the damage model on the material level. The paper concludes with a brief discussion of how this damage model may be applied to response analyses of concrete frames.

Introduction

Current seismic design philosophy relies strongly on the concept of energy dissipation through inelastic action. For the proper implementation of this concept it is essential that the structure, or rather selected structural members, be endowed with adequate ductility. This property, *ductility*, is such a fundamental pillar of earthquake-resistant design that it appears reasonable to periodically rethink its basic meaning and design implications.

In conventional design for static loads, the basic design requirement can be stated in the general form,

$$Design \ Strength \ \geq \ Load \ Effect$$

When designing for dynamic loads resulting from an earthquake, such a design requirement is typically supplemented with a corresponding ductility requirement,

$$\mu_{capacity} \geq \mu_{demand} \tag{1}$$

Use of a concrete member's or structure's ductility property in this simple format is incorrect, because it implies that both the ductility capacity and demand are independent of the number of load applications. It is known that this is not the case. In fact, the decrease of both strength and ductility with the number of load applications is well known and generally designated as fatigue. The number of load applications adds a second dimension to the problem. A more appropriate design requirement of the type of Eq. 1 would therefore take on the form,

$$\mu_{capacity}(N) \geq \mu_{demand}(N) \tag{2}$$

where N is the number of load applications. Eq. 2 is still an oversimplification, because in reality a structure is subjected to loads of varying amplitudes. In addition, the order in which a series of load impulses of different amplitudes are applied, has also an effect. For example, the damage resulting from one strong load impulse followed by a series of smaller loads can be substantially different from the damage resulting from the same strong load impulse preceded by the series of smaller loads.

In this paper, an energy-based model is proposed for the purpose of simulating the damage accumulation in concrete up to failure, due to a given sequence of load applications.

First, we shall discuss damage in general and introduce a definition of a suitable damage index. Because of the lack of experimental data on low-cycle fatigue of concrete, an investigation was initiated at Columbia University to study the damage accumulation of plain and fibre-reinforced concrete cubes in the low-cycle fatigue range. Some of the preliminary results are presented herein, together with a suggested model to numerically reproduce this behavior. The paper concludes with a discussion of how the concept of this energy-based damage model may be applied to detailed nonlinear dynamic analysis of reinforced concrete frames, by incorporating it into advanced time history analysis programs.

Definition of Damage

The concept of damage is familiar to everybody. Yet, if we have to define it in precise scientific terms, it appears that our notion of damage is in general rather vague. In the context of building response to destructive earthquakes, a large number of definitions have been proposed [1,6]. Most of these are of a more or less empirical nature and often prone to subjective influences. To avoid this problem, it is advantageous to tie the damage definition to the degree of physical deterioration with clearly defined consequences regarding the material's capacity to resist further load. Similarly, the notion of failure needs to be stripped of an occasionally arbitrary definition (on the member or structural level) by relating it to a specific level of damage at which the material or member ceases to resist further load. Herein we shall measure damage by an index D, equal to the ratio between the energy dissipated up to a certain point in time, E_i, and the total energy dissipation capacity, $\overline{E}_i$,

$$D = \frac{E_i}{\overline{E}_i} \qquad (3)$$

This definition can be substituted for the cycle ratio (n_i/N_{fi}) used in conventional fatigue analysis and is in general applicable both on a material and member level.

On the material level we are dealing with the stress-strain response of plain concrete. Failure is defined as the point at which the slope of the stress-strain curve ceases to remain positive. The energy dissipation capacity $\overline{E}_i$ per unit volume is clearly a material property, that can be measured in the laboratory as a function of stress or strain level i under constant-amplitude loading.

On the member level, for example, for a reinforced concrete frame element, the load-deformation or moment-curvature response is involved. Failure is now defined as the point at which the moment-curvature (or moment-rotation) curve ceases to have a positive slope, in the absence of an axial force. $\overline{E}_i$ on the member level can also be measured in the laboratory as a function of load or deformation amplitude i, but it obviously is a function of many other factors, such as reinforcement details.

Even though (or rather because) the above damage definition is very basic, it cannot easily be extended to the structure level, because on this level the loading has a spatial distribution component that is not present on the material or element level. For this reason, all so-called *global damage* definitions have an element of arbitrariness and lack the rational component that would be needed to evaluate a structure's residual capacity to resist further load. Therefore we shall not attempt to define damage on the structural (or global) level. It is not missed either, at least not in the context of nonlinear dynamic analysis of degrading structures.

It may be appropriate to note that a general definition of damage such as Eq. 3 lends itself to non-mechanistic sources of damage by referring to the remaining or residual energy dissipation capacity,

$$\overline{E}_i - E_i = \overline{E}_i(1 - D) \qquad (4)$$

For example, if a concrete specimen had been exposed to a certain number of freeze-thaw cycles or a given history of chemical attack, the inflicted damage D may be correlated to the remaining energy dissipation capacity, $\overline{E}_i - E_i$, which can be measured.

The most serious drawback of this damage definition is the lack of an experimental data base for $\overline{E}_i$. S-N curves have been generated for some selected cases [8], but for the low-cycle fatigue range that is of interest for earthquake response analysis, very few data exist.

On the member level, a considerable body of knowledge exists, thanks to extensive experimental investigations, that were carried out primarily at the University of California at Berkeley, the University of Illinois at Urbana, and the University of Michigan. However, because of the large number of influence factors it is difficult to organize the available data in a systematic way. The investigations which are most suitable for forming the foundation of the needed data base are those by Hwang [3] and Scribner and Wight [7].

Low-Cycle Fatigue of Plain and Fibre-Reinforced Concrete

In order to study the energy dissipation capacity of plain and fibre-reinforced concrete, 4-inch cubes were loaded cyclically under load control up to failure (for a detailed report see [2]). Table 1 summarizes some pertinent information on the specimens. All in all, nine different mixes were prepared, with typically fifteen specimens for each mix, three of which were loaded monotonically to failure, in order to determine the cube strength and to help select the various cyclic load amplitudes for the remaining cubes. Three different load amplitudes were tested, with three specimens for each data point. (The small number of samples for each data point underscores the preliminary nature of this investigation). The following stress ratios (i.e. applied stress divided by uniaxial cube strength) were selected: 0.75, 0.8, and 0.9. Several specimens were loaded with a stress ratio of 0.7 and required an extremely large number of cycles to failure. Since the objective of this study was the investigation of low-cycle fatigue behavior of concrete, this stress ratio was not pursued any further. At the upper limit, tests with stress ratio 0.9 and higher, because of statistical scatter, frequently led to failure during the first load cycle and thus had to be disregarded.

Table 1. Strength data for 4-in cube specimens

Specimen Mix	Fibre Content		Ave 28 day Cyl. Str. (psi)	Ave 28 day Young's Mod (ksi)	Ave 180 day 4 in Cube Str. (psi)
	Steel (%)	Polyprop. (%)			
PL			6,755	4,370	10,962
P0.25		0.25	6,720	4,420	9,835
P0.50		0.50	7,115	4,470	10,166
P0.75		0.75	6,490	4,400	10,357
P1.00		1.00	6,484	4,310	10,022
S0.25	0.25		7,663	4,660	10,367
S0.50	0.50		7,368	4,770	11,516
S0.75	0.75		8,105	5,020	12,052
S1.00	1.00		8,630	5,050	12,195

Typical load-deformation histograms are reproduced in Fig. 1. As can be seen, the tests were started with a small amount of prestress, below which the load was never lowered.

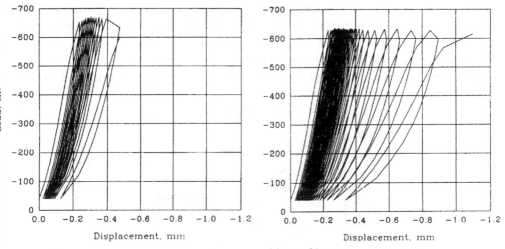

a) Plain Concrete
Number of Cycles to Failure, $N_f = 13$

b) 0.75% Polypropylene Fibre
Number of Cycles to Failure, $N_f = 42$

Fig. 1   Typical Load-Displacement Histograms

Table 2. Low-Cycle Fatigue Results

Specimen	Stress Level								
	0.9			0.8			0.75		
Plain	50	13	-	326	189	124	26,038	1,329	-
P0.25	920	441	409	9,068	5,822	5,444	161,982	4,765	18,100
P0.50	55	41	31	1,673	1,583	1,455	72,659	6,424	5,055
P0.75	59	42	30	636	205	188	3,566	699	190
P1.00	44	31	30	326	279	255	8,110	1,016	879
S0.25	1,676	112	94	2,886	737	248	5,432	1,804	1,182
S0.50	37	36	8	260	168	103	1,394	766	246
S0.75	15	11	10	153	152	33	1,005	719	404
S1.00	24	8	8	166	152	33	267	205	199

Table 2 gives for each specimen tested the number of cycles to failure. The statistical scatter is considerable but still reasonable in comparison with test results reported elsewhere in the literature. Graphs relating stress level to number of cycles to failure (S-N curves) are shown in Fig. 2, together with straight-line approximations determined by regression analysis. Since the dissipated energy was determined for each load cycle it was possible to draw the damage accumulation curves of Fig. 3, with damage being defined according to Eq. 3. These data clearly show that Miner's Rule, which presumes a linear accumulation of damage, is not generally applicable to either plain or fibre-reinforced concrete. Except for 0.25% volume of polypropylene fibre and a stress ratio of 0.75, all results exhibit a more or less pronounced nonlinear evolution of damage. The degree of nonlinearity increases almost consistently with stress ratio, regardless of volume and type of fibre. Damage initially accumulates at a slower rate and accelerates as the specimen approaches failure. For plain concrete, the damage accumulates nonlinearly for stress ratios in excess of 0.75, which implies that Miner's Rule may be applied with little error as long as stress ratios remain below 0.75.

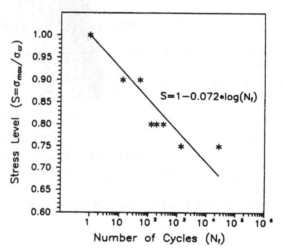

Fig. 2   S-N Curves for Plain Concrete

It should be restated that the data presented here are still preliminary in nature and based on a number of samples per data point that is clearly too small for a low-cycle fatigue investigation. Yet, a tentative damage accumulation function may be offered for plain concrete as follows:

$$
\begin{aligned}
D &= (1.7 - S)\, \bar{n} \quad &\text{for } \bar{n} \leq 0.6 \\
D &= \bar{n}^{1.6S} \quad &\text{for } \bar{n} > 0.6
\end{aligned}
\tag{5}
$$

where $\bar{n} = n_i/N_{fi}$ is the cycle ratio, and S = stress ratio.

For variable-amplitude loading, Eq. 5 may be used to convert the damage produced by n_1 cycles with some stress level S_1 to an equivalent number of cycles n_2^* with some stress level S_2 [4], by equating the corresponding damage levels D_1 and D_2^*,

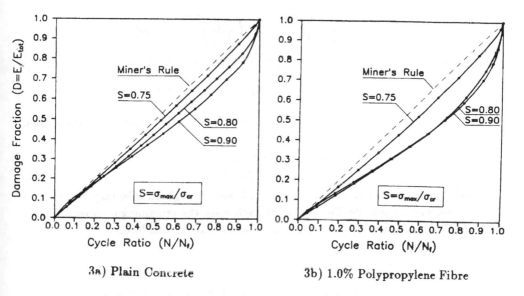

3a) Plain Concrete

3b) 1.0% Polypropylene Fibre

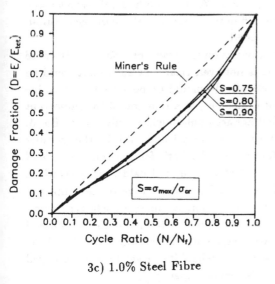

3c) 1.0% Steel Fibre

Fig. 3   Damage Accumulation Curves

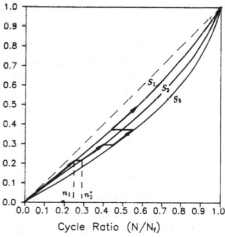

Fig. 4   Damage Accumulation for
Variable Load Amplitudes

$$D_1 = \left(\frac{n_1}{N_{f1}}\right)^{1.6S_1} = \left(\frac{n_2^*}{N_{f2}}\right)^{1.6S_2} = D_2^*$$

from which

$$n_2^* = \left(\frac{n_1}{N_{f1}}\right)^{\frac{S_1}{S_2}} N_{f2} \qquad \text{for } \bar{n} > 0.6$$

The combined damage produced by n_1 cycles of load level S_1 and n_2 cycles of load level S_2 is then,

$$D = \left(\frac{n_2^* + n_2}{N_{f2}}\right)^{1.6S_2}$$

Similarly, the evolution of damage can be followed through an arbitrary sequence of loads of variable amplitude, Fig. 4, and at each stage the remaining number of cycles can be determined that the material can resist for a given load level.

Low-Cycle Fatigue of Reinforced Concrete Members

For material characterization, linear elastic analyses and selected nonlinear analyses require only basic information as input, such as Young's modulus, Poisson's ratio, or the yield strength of steel. For degrading reinforced concrete, also damage accumulation information of the kind of Eq. 5 is needed.

On the member level, corresponding information is as yet not available. Because of the large number of influence factors, it is not even clear how such data could be organized, once they are available. Theoretically, it should be possible to determine the total energy dissipation capacity $\overline{E}_i$ of a member, as a function of the important influence factors and then to formulate a damage accumulation law equivalent to Eq. 5. The most significant parameters aside from the load amplitude i will be the member dimensions, concrete strength, and reinforcing details, both for flexural, shear and confinement reinforcement. The establishment of the required data base is clearly a formidable undertaking, but indispensible for a rational alternative to the empirical hysteretic load-deformation relationships with ever more complex Takeda-type rules. Only with such a damage accumulation law will it be possible to predict the response of a reinforced concrete member to an arbitrary load history all the way through failure. At the same time it will be possible to assess the reliability or safety of a damaged member by quantifying its residual energy dissipation capacity.

This analysis challenge is further complicated by the existence of different potential failure modes. If a member were known to fail in flexure, then a model could be constructed as described above. But shear and bond failure modes and concrete crushing cannot be ignored. Moreover, each failure mechanism has in general its own damage accumulation rate, so that it is virtually impossible to predict a priori which failure mode will eventually prevail.

Dynamic Analysis of Reinforced Concrete Frames

The basic objective of nonlinear dynamic analysis of a structure is to simulate numerically the response to load as realistically as possible. The member models incorporated in the various existing computer programs are either *empirical* or *rational*. In the *empirical* approach, experimental data are analyzed for the influence of selected dominant parameters, and some "law" is established with several free constants, which are determined typically by regression analysis or more sophisticated system identification techniques. An extreme example is the damage model of Park and Ang (1985),

$$D = \frac{\delta_{max}}{\delta_u} + \frac{\beta}{Q_y \delta_u} \int dE \qquad (6)$$

where δ_{max} = maximum deformation experienced so far, δ_u = ultimate deformation under monotonic loading, Q_y = calculated yield strength, dE = dissipated energy increment, and

$$\beta = \left(-0.447 + 0.73\frac{l}{d} + 0.24n_o + 0.314P_t\right) 0.7^{\rho_w} \qquad (7)$$

with l/d = shear span ratio, n_o = normalized axial force, ρ_w = confinement ratio, P_t = longitudinal steel ratio. The concatenation of the influence parameters deemed most significant in Eq. 7 has clearly an element of arbitrariness. Likewise, it is not obvious what physical evidence suggests the linear superposition of the two components that make up the damage index in Eq. 6. Therefore, it is uncertain whether the model yields realistic results outside the range of parameters for which it has been calibrated.

The *rational* approach towards constructing a model relies on basic material information of the kind described earlier and on physical insight and understanding of member response. The advantages are obvious in that the range of applicability is extended beyond a narrow domain of data used for calibration, and the element of arbitrariness is replaced by physical insight. The difficulties of the rational approach have been pointed out before and cannot be ignored. Yet, short of expensive experimental studies, there seems to be no alternative to fulfilling the analyst's objective of predicting the strength and stiffness degradation of reinforced concrete frames to a given load history and then using this information to provide rational assessments of the residual or remaining safety of a building that had been damaged by an earthquake.

Acknowledgment

The research described herein has been partially supported by the National Center for Earthquake Engineering Research through a grant from the U.S. National Science Foundation. The experimental investigation was conducted by Dr. M. Grzybowski of the Royal Institute of Technology, Stockholm, during a post-doctoral stay at Columbia University.

References

1. Chung, Y.S., C. Meyer, and M. Shinozuka, "Seismic Damage Assessment of Reinforced Concrete Members," Technical Report NCEER-87-0022, National Center for Earthquake Engineering Research, State University of New York, Buffalo, 1987.

2. Grzybowski, M. and C. Meyer, "Damage Prediction for Concrete With and Without Fibre Reinforcement", Dept. of Civil Engineering, Columbia University, New York, 1991.

3. Hwang, T.H., "Effects of Variation in Load History on Cyclic Response of Concrete Flexural Members", Ph.D. Thesis, Dept. of Civil Engineering, University of Illinois, Urbana, 1982.

4. Park, Y.J. and A.H.S. Ang, "Mechanistic Seismic Damage Model for Reinforced Concrete", Journal of Structural Engineering, ASCE, Vol. 111, No. 4, April, 1985.

5. Meyer, C. and M.S. Bang, "Damage of Plain Concrete as a Low-Cycle Fatigue Phenomenon", Dept. of Civil Engineering, Columbia University, New York, 1989.

6. Reitherman, R., "A Review of Earthquake Damage Estimation Methods," Earthquake Spectra, EERI, Vol. 1, No. 4, August, 1985.

7. Scribner, C.F. and J.K. Wight, "Delaying Shear Strength Decay in Reinforced Concrete Flexural Members Under Large Load Reversals, Report No. UMEE 78R2, Dept. of Civil Engineering, University of Michigan, May 1978.

8. Shah, S.P., ed., "Fatigue of Concrete Structures", ACI Special Publication SP-75, ACI, Detroit, 1982.

SEISMIC MOTION DAMAGE POTENTIAL FOR R/C WALL-STIFFENED BUILDINGS

K. MESKOURIS, W.B. KRÄTZIG, U. HANSKÖTTER
Lehrstuhl für Statik und Dynamik
Ruhr-University Bochum
P.O. Box 102148, D-4630 Bochum 1, Germany

ABSTRACT

Seismic damage in buildings is heavily dependent on the number of load reversals experienced, which correlates mainly with strong motion duration. While for sites with low seismic damage potential linear models are sufficiently accurate for design purposes, nonlinear models are essential if a large number of load reversals is to be expected. A classification of probable ground motions at a site according to their damage potential is proposed and some accompanying problems are discussed in connection with nonlinear analyses of large wall-stiffened R/C buildings.

INTRODUCTION

It is well-known that shear walls greatly reduce seismic damage in high-rise buildings by drastically limiting drift demand. If we are dealing with a site of high seismic motion damage potential, implying a large number of load reversals, it is meaningful, at least for buildings of a certain importance, to conduct nonlinear analyses which furnish the structure's temporal damage evolution, in order to be able to assess already in the design stage merits and weaknesses of a certain design variant.

Description of seismic risk at a site by a single parameter, e.g. peak ground acceleration, PGA, is not very meaningful, and it has been observed that damage does not correlate well with measured PGA values. To quote HOUSNER, who, in his monogragh [1, p.57] has drastically illustrated this problem, "...peak acceleration gives a completely misleading impression of the destructive potential of accelerograms like Melendy ranch and Rocca, and seriously underestimates the strength of artificial records...which have most of their energies in lower frequencies than those that typically determine peak acceleration". On the other hand, peak ground velocity, velocity spectral ordinates and strong motion duration values (the latter defined after [2] as time elapsed between 5% and 95% of the HUSID-diagram) are better descriptors, being directly connected with the energy involved. FAJFAR [3] proposed a factor $I=v_g\,t_D^{0.25}$ as an intensity measure for scaling ground motions, with maximum ground velocity v_g and strong motion duration t_D. Some problems associated with such intensity measures will be briefly addressed in the next section before considering the problem of nonlinear analysis of wall-stiffened R/C buildings.

DAMAGE POTENTIAL OF GROUND MOTIONS

Regarding damage potential of ground motions, a classification of probable records into one of three types (S, M and L) has been proposed [4] according to characteristics such as strong motion duration, central period and quotient of max. ground acceleration to max. ground velocity (see Table 1 below).

TABLE 1

Proposed classification of ground motions

	S-type	M-type	L-type
- Duration t_D [s]	< 10	$10 < t_D < 15$	> 15
- Central period T_0 [s]	< 1	$1 < T_0 < 1.2$	> 1.2
- a_g/v_g [g/m/s]	> 1.0	0.8 to 1.0	< 0.8

L-type motions (low frequency, long duration, high energy) are characterized by a great number of load reversals and accordingly high demands on nonlinear reserves, while S-type records (high frequency, short duration, low energy) do not pose as serious a threat to structural integrity. While for S- and M- type excitations usual response-spectrum based design and analysis techniques in conjuction with detailing requirements are sufficient, the damage evolution of important buildings subject to L-type excitations should be investigated by nonlinear Direct Integration. To illustrate, Figs. 1 and 2 compare linear and nonlinear time histories of the ground floor drift of an R/C frame subjected to scaled versions of the SCT-N90W Mexico City 1985 acce-lerogram (L-type) and the ground motion recorded at Petrovac during the Mon-tenegro 1979 event (M-type). The accelerograms were scaled so that they had identical spectral ordinates at the fundamental period of the structure. It follows that for the linear model (Fig. 1), drift maxima were the same for both motions, while in

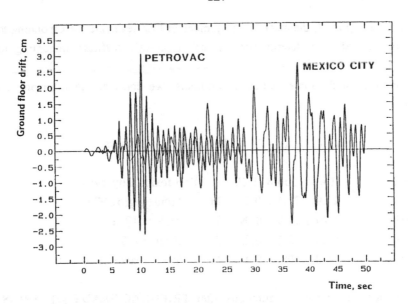

Figure 1. Ground floor drift for two motions, linear model.

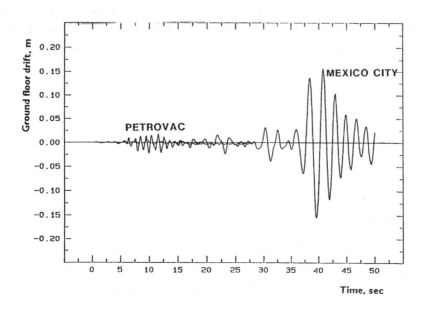

Figure 2. Ground floor drift for two motions, nonlinear model.

the nonlinear case (Fig. 2) the much larger number of load reversals corresponding to the SCT record led to a deterioration of the structure's stiffness amounting to collapse.

In order to study some particular problems, we consider the following five measured records:

ID	t_D	$I_{AR.}$	$I_{FAJF.}$	T_0	Type	Remarks
SCT	38.8	15.2	150.8	1.60	L	Mexico City 1985
PET	10.6	27.9	72.1	0.27	M	Montenegro 1979
ROC	3.03	2.70	14.5	0.09	S	Ancona 1972
TOL	4.31	5.05	32.1	0.22	S	Friuli 1976
IGN	4.96	14.0	108.5	0.41	M	San Salvador 1986

Given above are strong motions durations after TRIFUNAC/BRADY [2], ARIAS intensities defined as time integrals of acceleration squared over the entire record duration given in $[m^2 s^{-3}]$, intensities after FAJFAR [3] in $[cms^{-3/4}]$, central periods defined as the reciprocal of the mean number of positive zero crossings per second and finally the record types according to the classification in Table 1. Fig. 3 shows acceleration response spectra for all five records for a damping ratio of 5% critical. The similarities between spectra of records of the same type are obvious.

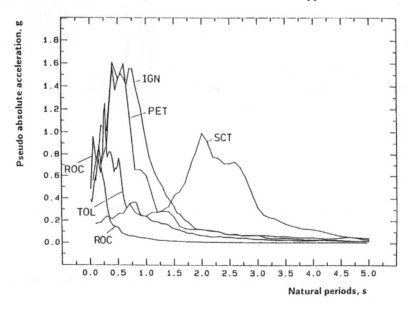

Figure 3. Acceleration response spectra D=5%.

A word of caution is in order regarding the determination of these values for L-type records. In the course of determining ARIAS intensities, strong-motion durations and other characteristics of measured records, a portion of the long "tail" (see Fig. 4) is often arbitrarily disregarded as being of no importance. This can, however, lead to quite noticeable errors. To demonstrate, the SCT N90W record has a strong motion duration of 29.1 s and an ARIAS intensity of 14.7 m^2s^{-3} for a nominal duration of 70s, while for a nominal duration of 160s the corresponding values are $t_D = 38.8$ s and I = 15.2 m^2s^{-3}. Generally, ARIAS intensities are not as sensitive to such cutoffs as strong motion durations. Fig. 5 shows values of computed strong motion duration of the SCT N90W record as a function of the number of time steps considered (each time step equal to 0.02s), together with the corresponding regression line. Obviously, for S- and M-type records no such problems exist.

If we consider nonlinear effects of S-, M- and L-type motions by computing nonlinear response spectra or evaluating other damage indicators over wide period ranges the following picture emerges: For slight nonlinearities, there is no marked difference in nonlinear responses to L- and M-type records, as can be seen in Fig. 6 which shows maximum ductilities for the three records SCT, IGN and PET in the period range up to 3 s. Maximum ductilities, defined as maximum elastoplastic deformations u_{max} divided by the limiting elastic deformation u_{el} were calculated in Fig. 6 for u_{el} values equal to 90% of the corresponding linear spectral displacement ordinates S_d. However, for higher ductility levels L-type records produce much higher ductility demands, as demonstrated in Fig. 7 showing maximum ductilities for elastic limits of 50% S_d. A non-degrading bilinear hysteresis model was utilized for these calculations, while with degrading TAKEDA-type models the difference is even more pronounced. S-type records exhibit a similar behavior to M-type accelerograms, but will generally produce none or at the most very limited inelastic action in well-designed structures.

For sites where only S-type ground motions are expected to occur, design and analysis of structures can proceed along well-established linear lines, even utilizing quasistatic methods. An approximate ultimate load analysis technique [5] has been employed by the authors for quantitatively evaluating ductility demands in framed structures, which yields reasonably accurate results as long as no significant strength and stiffness degradation takes place, that is for S- and M-type motions. We feel however that for sites with possible L-type motions and also for structures of special importance there exists currently no alternative to step-by-step dynamic nonlinear analyses for dependably evaluating nonlinear demand and checking for the presence of weak points which may drastically reduce the overall safety level. Of course, the rough classification of Table 1 for S-, M- and L-motions is open to further refinement,

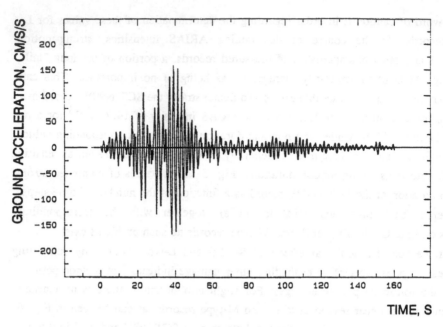

Figure 4. SCT N90W accelerogram.

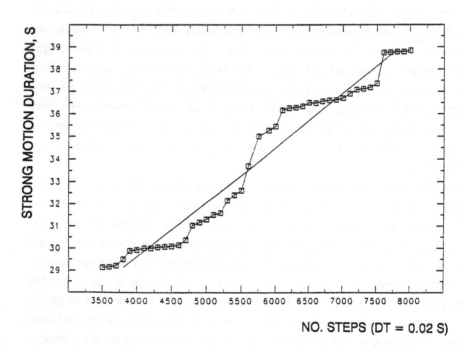

Figure 5. Duration as function of record length, SCT N90W

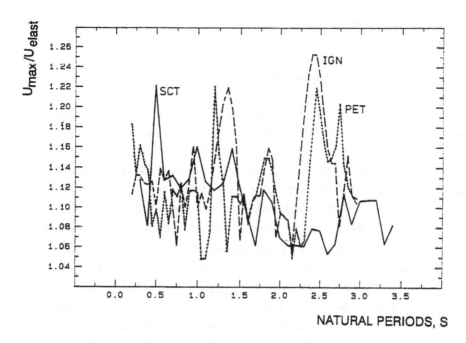

Figure 6. Maximum ductilities for $u_{el} = 0.9\ S_d$.

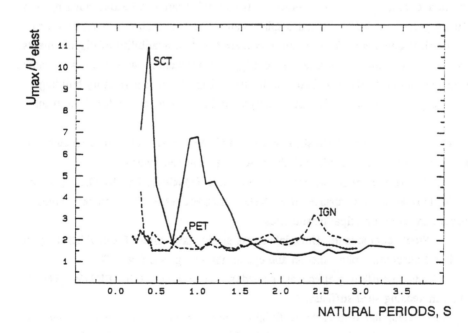

Figure 7. Maximum ductilities for $u_{el} = 0.5\ S_d$.

especially on the basis of energy measures in conjuction with special types of buildings.

WALL-STIFFENED R/C BUILDINGS

In contrast to buildings stiffened exclusively by moment-resisting frames, where story drifts exhibit large variations, shear walls acting as "stiff backbones" ensure a much more uniform distribution and also a drastic limitation of story drifts. A problem associated with this drift limitation is, however, a higher risk of non-ductile failure modes due to shear. As stated in the 1985 CEB Model Code for Seismic Design of Concrete Structures, maximum shear forces in a structural wall with given moment capacity can considerably surpass the values predicted by linear analysis. While much useful information can be extracted from nonlinear analyses of simple beam-like shear-wall models, full three-dimensional computer models are needed if one is to closely monitor damage evolution under given excitations. Such a computer program [6] for the nonlinear seismic analysis of wall-stiffened three-dimensional R/C buildings has been developed in the last years at Bochum. It is embedded in the program system FEMAS [7] and is being continuously expanded and applied to the analysis of real-life buildings. The standard idealization of rigid floors in their own plane is adopted, so that lateral movements of each floor can be described by two mutually perpendicular horizontal displacements and the rotation around a vertical axis. Vertical degrees of freedom are considered and compatibility of displacements at vertical wall joints is enforced, so outrigger constructions with their well-known advantages can be easily modelled. To illustrate, Fig. 8 shows a 16-story building and its first eigenvector, which has been calculated in order to check for discretization errors.

For the idealization of the shear walls, a refined isoparametric R/C rectangular layered element is being used, with the following characteristics:

- Concrete component: Material law after DARWIN/PECKNOLD [8], see Fig. 9, biaxial failure surface after KUPFER/GERSTLE [9], concrete cracking criterium based on principal tensile stress.

- Steel component: Uniaxial stress-strain law after PARK/KENT [10], including Bauschinger phenomenon and cyclic hardening (see Fig. 10).

- Tension stiffening effects are considered according to GILBERT/WARNER [11] by increasing steel stiffness.

A layered rectangular isoparametric finite element for reinforced concrete based on these assumptions has been developed in [6] and is employed for the nonlinear analysis. For real-life situations, with many structural walls and over dozens of stories,

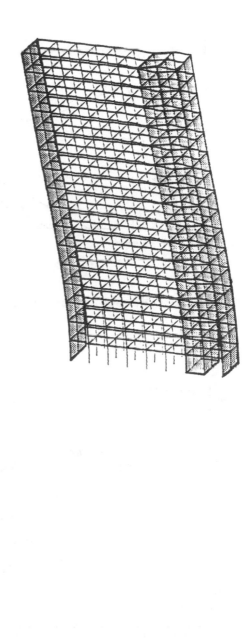

Figure 8. 16-story building and fundamental eigenvector.

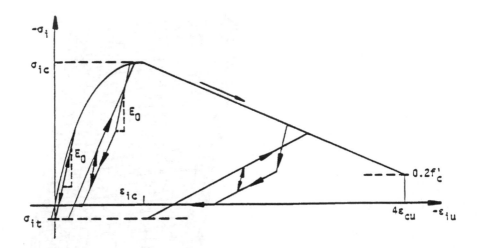

Figure 9. Concrete constitutive law after DARWIN/PECKNOLD.

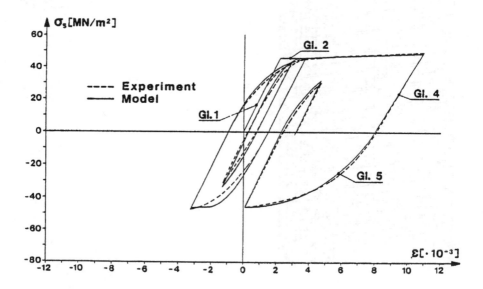

Figure 10. Steel constitutive law.

the working load becomes prohibitively large even for modern large computer workstations, so every possibility for reducing overall effort has been investigated. Such techniques include

    - Automatic generation of input data by means of special preprocessors requiring minimal input, automatic optimization of bandwidths,

    - Static condensation for suitably chosen master degrees of freedom,

    - Modified NEWTON/RAPHSON solution during time integration by utilizing a diagonal effective stiffness matrix; this dispenses with the need for recurring triangularization and is quite time effective even if more iterations to convergence are required than with the usual banded matrix.

A special technique has been further developed [12] for coding complex material laws by using graph theory and expert system techniques. It eliminates logical mistakes in this extremely error-prone area and ensures simple interchange of different material models. Currently, the influence of partition walls etc. which nominally do not play a part in distributing lateral loads, but which do, in fact, influence overall load-carrying capacity, is being investigated by modelling them as nonlinear spring units with empirical load-deflection characteristics.

RESULTS AND CONCLUSIONS

Results obtained confirm the beneficial reduction of drift by utilizing R/C walls as stiffening elements for lateral loads. Consideration of vertical flexibility and compatibility of vertical deformations has turned out to play quite an important role for the correct modelling of system behavior. In the frame of such a computer model it is also possible to take into account the temporal variability of earthquake direction and also, of course, of the ground motion's vertical component. As of now, we have not, however, studied these effects in any depth.

To summarize:

a) A classification scheme is proposed for describing the damage potential associated with probable ground motions at a site. This scheme can be extended to include energy variable based criteria for special structures, which determine the degree of sophistication of the analyses required.

b) For buildings of some importance which could be exposed to L-type seismic excitations with high numbers of load reversals, nonlinear dynamic analyses are advocated already in the design stage in order to assess probable structural behavior and damage evolution.

c) A computer program has been developed for this task which conducts the investigation of three dimensional reinforced concrete buildings stiffened with structural walls as well as with frames. Special techniques were necessary for reducing work loads in data preparation, nonlinear analysis and post-processing of results in order to stay below prohibitive levels.

ACKNOWLEDGMENTS

The continuous financial support of the "Deutsche Forschungsgemeinschaft (DFG)" is gratefully acknowledged.

REFERENCES

1. Housner, G.W and Jennings, P.C, Earthquake Design Criteria. EERI Monograph Series, Earthquake Engineering Research Institute, Berkeley, California, 1982.

2. Trifunac, M.D. and Brady, A.G., A Study on the Duration of Strong Earthquake Ground Motion, Bulletin of the Seismological Society of America, Vol. 65, No. 3, June 1975, pp. 581-626.

3. Fajfar, P., Vidic, T., Fischinger, M., A measure of earthquake motion capacity to damage medium-period structures, Soil Dynamics and Earthquake Engineering 1990, Vol. 9, No. 5, pp. 236-242.

4. Meskouris, K. and Krätzig, W.B., Seismic damage assessment of buildings. Proceedings, ERCAD Berlin 1989, Balkema-Verlag 1990, pp. 427-433.

5. Krätzig, W.B., Meskouris, K.: Nachweiskonzepte für erdbebensichere Ingenieurtragwerke. Tagungsheft "Baustatik-Baupraxis 4", Hannover 1990.

6. Elenas, A., Ein Beitrag zur physikalisch nichtlinearen Analyse erdbebenerregter räumlicher Aussteifungssysteme aus Stahlbetonscheiben. Diss. Bochum 1990.

7. Beem, H. et al., FEMAS, Finite Element Moduln Allgemeiner Strukturen, Benutzerhandbuch Version 90.1, Institut für Statik und Dynamik, Ruhr-Universität Bochum 1991.

8. Darwin, D. and Pecknold, D.A., Nonlinear Biaxial Law for Concrete, J. Eng. Mech. Division, ASCE, Vol. 105 (1979), p.623.

9. Kupfer, H.B. and Gerstle, K.H., Behavior of Concrete under Biaxial Stresses, J. Eng. Mech. Division, ASCE, Vol. 99 (1973), p. 852.

10. Park, R., Kent, D.C., Sampson, R.A., Reinforced Concrete Members with Cyclic Loading, J. Struct. Division, ASCE, Vol. 98 (1972), p. 1341.

11. Gilbert, R.I., Warner, R.F., Tension Stiffening in Reinforced Concrete Slabs, J. Struct. Div., ASCE, Vol.105 (1978), p. 1885.

12. Hanskötter, U. et al., A knowledge-based approach to the numerical treatment of constitutive laws for cyclic loading. Structural Dynamics, Proceedings EURODYN 1990, Balkema-Verlag, Vol.1, pp.465-470.

COMPARISON OF DAMAGE MEASURES FOR THE DESIGN OF REINFORCED CONCRETE FLEXURAL MEMBERS

Christos A. Zeris
Research Engineer, Reinforced Concrete Laboratory
National Technical University of Athens,
42 Patisson Str., Athens GR 106 82, GREECE

ABSTRACT

The use of refined element formulations for the modeling of flexural behavior of reinforced concrete columns allows for the evaluation and comparison of different indices used to quantify flexurally induced damage. The results of such a case study are presented considering biaxial monotonic and cyclic excitations. Such spatial excitations are known to induce increased amounts of damage compared to uniaxial response. Indices considered include conventionally used displacement ductilities as well as the absorbed hysteretic energy, both along the member and across the critical section.

INTRODUCTION

Several methods quantifying the damage potential of earthquakes have utilized maximum, cumulative or cyclic lateral drift, either absolute or normalized as ductility. Ductility limits have traditionally been used to examine the ability of a system to undergo inelastic response for a given base input, leading to the utilization of specified ductility values for design. The sensitivity of this measure to different modeling conventions, base input content and normalization, or the integration procedure, has been considered in detail by several investigations (e. g. Fajfar and Fischinger 1984, Mahin and Lin 1983). The reliability of this design approach for single mass systems has been discussed by Mahin and Bertero (1981).

More recent investigations consider the input and absorbed hysteretic energy of single degree of freedom oscillators as a measure of damage under seismic type loading (Akiyama 1985, Uang and Bertero 1988). Energy input to the oscillator is found to be a fairly stable quantity for base excitation normalizations and is therefore pursued as a characteristic damage index for single degree of freedom analyses and possible use in design response spectra (Fajfar et al. 1989). Energy absorption can account also for factors such as the number of reversals and duration and represents therefore an improved parameter to characterize failure of the structural system.

Spatial response has also been studied at this single mass level, in order to account for more realistic excitation and response conditions; different hysteretic and interaction rules, stochastic or deterministic excitation and variable angle of incidence effects have been considered by Nigam (1967), Takizawa and Aoyama (1967), Park et al. (1986) and Zeris (1990). Generally, it is observed that the total energy absorbed into a system is constant with angle of incidence of the earthquake, however, the ratio of energy absorbed among the two orthogonal directions is not. In fact, considerable variations have been found to occur for systems of short period or widely different normalized strengths in each direction even under the same base input but varying angle of earthquake incidence.

Analytical efforts currently attempt to correlate such global measures to local behavior through multi degreee of freedom analyses; most emphasis is yet placed into the relation between global drift ductilities and local damage, since such an index is traditionally used in design codes.

At the member level, a combination of energy ductility and normalized deformation (rotational or curvature) is favoured as an index of damage quantification. Proposed member models account for local mechanisms that govern the response of reinforced concrete members and are calibrated by experimental evidence. The model proposed by Park et al. (1985) uses a combination of the two indices above, namely absorbed energy and deformation. A different model is proposed by Chung et al. (1989) that elaborates on the representation of history to failure by accounting, in addition, for cyclic history and the number of cycles. The latter model uses a more realistic representation of the distribution of damage within the member but does not account for axial load variation and biaxial excitation influence.

Behaviour of reinforced concrete structures and members (Oliva 1980, Otani and Cheung 1982, Low and Moehle 1986, Shahrooz and Moehle 1987) and investigation of reinforced concrete structures (Zeris 1986) in spatial response, have indicated increased damage compared to planar only response. Observation and analysis indicate that biaxial effects complicate the hinge distribution and therefore energy absorption in the structure. Futhermore, increased flexural damage in columns is evidenced locally by spalling the cover concrete, inelastic straining and fixed end pullout in tension, or buckling in compression of the reinforcement. Hence, improved local damage calibration is necessary in the presence of generalized spatial response, particularly for columns. The purpose of this investigation is to examine the reliability of the conventional indices of drift and energy ductility to depict such local damage, induced under biaxial response, and to quantify such damage using the conventional measures previously described.

A refined finite element model is used that accounts for spread inelastic action within the column under generalized axial/flexural excitation where shear associated phenomena are not significant. So as not to complicate the parameters involved, only contributions from spread damage are considered herein; fixed end effects can be considered by using additional inelastic representations of the pullout mechanisms at the column joint, implemented with this model by Zeris (1986). Limited parametric studies of this relative contribution have been reported by Zeris (1986) and are beyond the scope of this work.

ESTIMATION OF DAMAGE INDICES OF A REINFORCED CONCRETE FLEXURAL COLUMN

Description of the model

A reinforced concrete cantilever column 3.0 m tall is analyzed under different tip excitations and constant axial loads. The column section is 50 cm by 50 cm square, 3 cm cover, reinforced with twelve 20 mm diameter bars uniformly spaced along the perimeter. Two different axial loads are considered corresponding to normalized axial force ratios v (namely load normalized by $A_g f_c$), of 0.1 and 0.2, where A_g is the gross column area of the section and f_c the concrete compressive strength.

Steel mechanical characteristics are modeled with an explicit cyclic hysteretic rule (Menegotto and Pinto 1977) following a trilinear virgin loading curve, with a yield strength of 400 MPa, elastic modulus of 200000 MPa and a hardening slope of 60000 MPa. Buckling in compression is not considered. Different uniaxial stress-strain characteristics are assumed for the core and cover concretes: both have similar ultimate compressive strength (f_c) of 20 MPa at 0.2% strain, and no tensile resistance. Cover concrete provides no compressive resistance beyond a compressive strain of 0.35%. The core exhibits a softening resistance past the maximum strength to a compressive strain of 2% at 20% f_c.

Each column section is modeled using about 120 concrete and 12 steel fibers. In all, four sections are monitored along the length, unequally spaced at 0.50, 1.00 and 1.50 m apart; use of flexibility interpolation functions enables a smaller amount of monitored sections than other finite element formulations. Two sections are monitored at the ends with full equilibrium compatibility between external and internal (section) forces; details of the model and algorithms are given elsewhere (Zeris 1986, Zeris and Mahin 1991).

Analytical investigations

The column is subjected to imposed monotonic or cyclic tip displacements patterns. Results of each inelastic analysis include load - tip deformation response, section level moment - curvature behavior along the member length and steel stress-strain characteristics across the base monitored section. Of particular interest is the distribution of inelastic flexural energy along the member and the strain energy of the section reinforcement, that primarily controls the desired ductile response of such members. Both the section flexural energy (E_f) and energy absorbed through internal straining of the reinforcement (E_s) are expressed as ductilities being divided by the corresponding energy at yield.

At the reinforcement level, strain energy ductility μ_{Es} is defined in terms of the yield strength f_y and modulus E, following Eq. 1:

$$\mu_{Es} = \frac{E_s}{E_y}; \qquad E_y = \frac{1}{2}\frac{f_y^2}{E} \tag{1}$$

while, due to the biaxial nature of the problem, section flexural energy ductility μ_{Ef} (per unit length) is defined in each direction as

$$\mu_{Ef} = \frac{E_f}{E_{fy}^{x,y}} \;\; ; \qquad E_{fy}^{x,y} = \frac{1}{2} M_y(\Theta)^{x,y} \, \Phi_y(\Theta)^{x,y} \tag{2}$$

where E_f is the absorbed flexural energy per unit length and $M_y^{x,y}$, $\Phi_y^{x,y}$ the projected x and y direction moment and corresponding section curvature at the onset of yield of the reinforcement anywhere in the section; the above quantities change with the angle of obliqueness of the tip response Θ. The axes x any y correspond to the principal axes of the member.

Analysis of Monotonic Behavior

Global characteristics. The cantilever is subjected to a monotonically increasing tip deformation to a maximum vectorial magnitude equal to 1.5 % of the height of the member. Different angles of incidence (kept constant) are considered, namely 0°, 30° and 45°. Since the member undergoes the same vectorial tip deformation, maximum projected drifts along each axis are reduced along x as Θ increases from 0°, following trigonometric ratios.

The projected moment-chord rotation (tip deformation over height) characteristics in each principal direction and the base moment paths in the M_x - M_y plane are compared in Figs. 1 and 2, for the two axial loads considered. The comparison of global characteristics indicates an expected reduction of ultimate flexural resistance under increasing, following biaxial interaction. For the same angle Θ, the strength in each direction increases with axial load as all cases considered are below the balanced load. The predicted yield and ultimate moments in each direction with corresponding chord rotations (namely tip deformation u normalized by the cantilever height L) are given in Table 1.

For improved quantification of damage in biaxial response, vectorial quantities are also considered in Table 1; vector deformation has been used at the single mass idealization of dynamical systems as a response index under spatial excitation. Due to the fact that a constant incidence Θ is imposed herein, the onset of yield chord deformation is well defined for the entire excitation history and is used in ductility estimations.

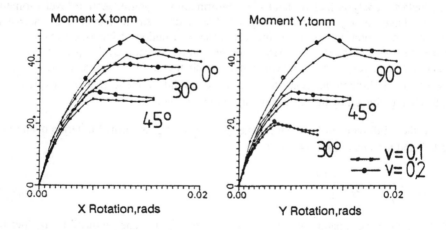

Figure 1. Moment-chord rotation characteristics

The vector sum of projected resistances at yield (Table 1) is only slightly dependent on Θ since the capacity surface in moment space at fixed axial load is circular for a square column. However, the vector displacement at yield (u_y) changes with Θ since yielding of the tension corner steel at the base occurs under oblique incidence; this affects somewhat the secant stiffness of the column, as shown in Fig. 1.

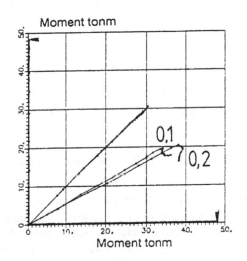

Figure 2. Base moment path under imposed constant drift

Table 1
(Moments are given in kNm and rotations in rad/100)
(a) Projected global level characteristics

Θ	v	M_y^x	M_y^y	u_y^x/L	u_y^y/L	M_u^x	M_u^y	$u^x(M_u)/L$	$u^y(M_u/L)$
0		336	-	0.55	-	425	-	1.10	
30	0.1	262	150	0.37	0.21	341	198	0.65	0.37
45		201	201	0.29	0.29	280	280	0.49	0.49
0		418	-	0.60	-	482	-	0.85	-
30	0.2	307	166	0.41	0.24	381	200	0.78	0.45
45		248	248	0.33	0.33	304	304	0.57	0.57

(b) Vectorial quantities, global level

Θ	v	M_y	u_y/L	M_u	$u(M_u)/L$	u_{max}/L	μ_u
0		336	0.55	425	1.10	1.50	2.7
30	0.1	302	0.42	394	0.74	1.50	3.6
45		284	0.41	396	0.70	1.50	3.6
0		418	0.60	482	0.85	1.50	2.5
30	0.2	349	0.47	436	0.90	1.50	3.2
45		351	0.47	430	0.81	1.50	3.2

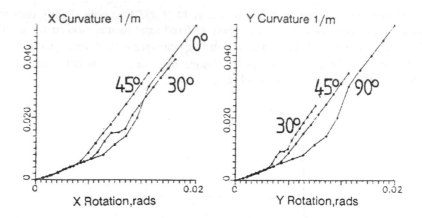

Figure 3. Variation of base curvature with chord rotation

Due to earlier onset of yield, the vector drift ductility demand (μ_u) increases with Θ even if drift magnitude is constant. In fact, projected ductilities (namely drift projections about each axis normalized by the uniaxial tip displacement at yield along this axis) are reduced as Θ increases to 45^0 by the cosine of Θ, and therefore, cannot reliably predict the actual behavior, even if summed vectorially, following the proposed expression

$$\mu = (\mu_x^2 + \mu_y^2)^{1/2} \tag{3}$$

The extent of local flexural damage is further elaborated in the following sections where inelastic strain energy and absolute reinforcement strain demands across the member sections are considered in some detail.

Despite the constant drift magnitude imposed, some deformation softening due to cover spalling is evidenced with increasing Θ that is not observed at 0^0. Assuming that the ultimate usable global deformation of the member occurs at a strength loss of 20 % it is evident from Fig. 1 that the column can adequately resist the imposed deformation in all cases but the 30^0 obliqueness where such a criterion is violated. Such is not the case under uniaxial bending alone. This preferential response can be also visualized by the resistance path beyond the maximum capacity surface in the moment paths of Fig. 2: the path direction at 45^0 remains, due to section symmetry, always on the diagonal while, in the case of 30^0 it diverges soon after maximum resistance is reached away from the x axis; this is attributed to the relatively faster reduction of the bending lever arm for bending due to tip movement along the y axis.

Section characteristics. The projected moment-curvature diagrams obtained at the base section of the cantilever are compared in Fig. 4, while the variation of base curvature with imposed chord rotation in each direction is given in Fig. 3 (for clarity, only the case of v equal to 0.2 is shown). The two figures suggest that the rate of increase of base curvatures

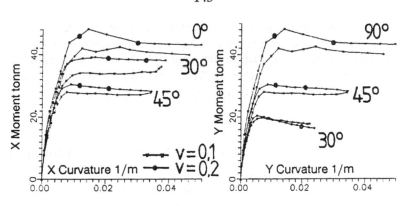

Figure 4. Base section moment-curvature characteristics

past yield is dependent on Θ and axial load, particularly for small angles of oblique deformation away from the axis of symmetry of the cross section.

Early cracking of the member base section under constant axial load reduces considerably the flexural stiffness of the column under intermediate angles of incidence. Maximum demanded curvatures and corresponding curvature ductilities at the base section are given in Table 2.

TABLE 2
Projected characteristics at the base critical section

Θ	v	Φ_y^x 1/m	Φ_y^y 1/m	Φ_{max}^x 1/m	Φ_{max}^y 1/m	μ_Φ
0		0.0067	-	0.046	-	6.8
30	0.1	0.0049	0.0028	0.037	0.022	7.6
45		0.0039	0.0039	0.034	0.034	8.8
0		0.0077	-	0.045	-	5.8
30	0.2	0.0054	0.0031	0.039	0.024	7.2
45		0.0042	0.0042	0.034	0.034	8.2

Similarly to the chord rotations, maximum demanded curvatures follow the trigonometric ratios in the two orthogonal directions, related by the inclination; as seen in Fig. 2 the column moment path deviates from this line in all but the 0^0 case, with certain dependence on axial load. Unlike the uniaxial response, maximum curvature ductility demands depend on the axial load v and on Θ, with the 45^0 obliqueness inducing the maximum demand, at v of 0.1.

The predicted flexural and reinforcement strain energy ductilities μ_{Ef} and μ_{Es} are summarized in Table 3. A possible measure of the relative damage along the member can be the ratio of flexural energy per unit length, absorbed at the base of the cantilever (E_{f1}) and, at a section one depth away (E_{f2}, Table 3). This ratio has a minimum value of twelve, uniaxially, increasing to 16 or 18 at 45^0 depending on the axial load.

TABLE 3
Absorbed energy and corresponding ductilities at the section and reinforcement level

Θ	v	ΣE_f kNm/m	μ_{Ef} x/y	E_{f1}/E_{f2} x/y	max μ_{Es}	S μ_{Es}	max ε_s
0		17.4	15	12	13.8	88	0.015
30	0.1	17.5	19/18	16/13	16.4	79	0.017
45		15.7	22/22	16/16	18.7	87	0.019
max/min			(1.44)		(1.35)		(1.31)
0		17.5	13	13	13.0	91	0.014
30	0.2	16.8	16/17	15/15	15.5	79	0.016
45		17.7	18/18	18/18	16.3	82	0.016
max/min			(1.41)		(1.25)		(1.14)

Similarly, the relative localization of damage anywhere across the base section is obtained from the maximum strain energy ductility absorbed by a single bar (tipically at the corner), or its maximum strain demand for the excitation at hand. Finally, the total strain energy absorbed by all the bars of the base section serves also to compare the relative extent of damage among the different deformation patterns considered.

The comparison of the analytical results indicates that the total flexural energy absorbed at the critical section in both directions (Eq. 2) is rather insensitive to the obliqueness (as also noted by Akiyama (1985)). Strong sensitivity with Θ however is observed for the projected flexural ductility, maximum strain energy ductility and maximum strain demand of the extreme tension reinforcement, which indicate increased damage away from uniaxial response. According to these indices, the 45^0 exhibits the highest relative increase of damage for both axial loads.

On the other hand, the total strain energy absorbed by the section steel (Table 3) is also only slightly affected by Θ, verifying the experimental observation that the corner steel contributes more to the total energy absorbed during oblique motion. Biaxial response predictions of absorbed strain energy at the section level (E_s), unlike global drift indices, considered earlier, do corroborate better the well established observation of increased local damage in biaxial, compared to uniaxial behavior.

Furthermore, considering the ratio E_{f1}/E_{f2} it is shown that the interior sections' contribution to bending energy absorption drops as Θ increases. The increase in axial load augments this phenomenon by controlling the ratio of base ultimate to yield moments along the member, and therefore the flexibility distribution. Such a tendency may not be possible to predict with the constant plastic hinge assumption often adopted in local damage characterization of more simplified member models.

Analysis of Cyclic Behavior

The member is reanalyzed being subjected to cyclic tip deformation between the same vectorial magnitude extremes as above but in spatial excitation, in order to extend the above observations to generalized response. Only the case of axial load of 0.2 A_g f_c is considered. The patterns of tip response are shown in Fig. 5 below. The first pattern involves four semi excursions at 1.5% drift along a plane (cases of displacement obliqueness of 0^0, 30^0 and 45^0 are considered), while the second pattern involves the same motion along x with intervening excursions along y; in this case, vector drifts are 40% higher than in the plane oblique response.

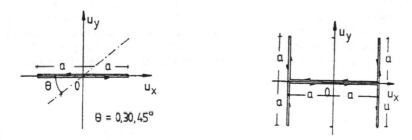

Figure 5. Biaxial deformation patterns considered.

The latter pattern of deformation (denoted double T) simulates the biaxial response of the ground corner columns of an actual structure which was damaged during the 1970 Imperial Valley earthquake (Zeris 1986); however, the analysis of the members of that specific structure is beyond the scope of this work. This type of deformation pattern was induced by the structural system used, namely a flexible ground storey moment frame longitudinally and a stiff wall-frame system in the transverse direction. The combination of these two systems, as well as base motion characteristics forced the column to respond in the frame direction over relatively large drifts while undergoing multiple significant inelastic

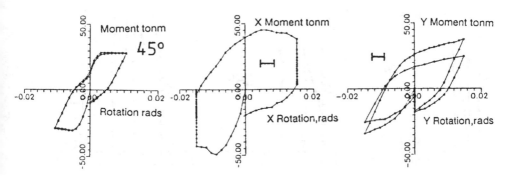

Figure 6. Moment-chord rotation under cyclic response

reversals in the orthogonal direction.

The global moment-chord rotation characteristics of the column about the two axes are given in Fig. 6. Maximum resistance, stiffness, hysteretic response shape and post ultimate strength degradation vary with the angle of incidence and the pattern of deformation. For the case of the double T, the column exhibits a strong degradation of strength and stiffness in the x direction, due to the intervening cycling along y: intermediate cycles strain the corner reinforcement beyond levels comparable to those under planar oblique response, modifying completely the resisting mechanisms in the transverse direction.

The consequence of such intermediate cycles is a significant increase of the strain energy demanded by the corner steels, compared both to the previous uniaxial response and to the oblique plane biaxial behavior herein, due to accumulated inelastic straining (Fig. 7). On the other hand, as shown in Fig. 8, the total absorbed flexural energy at the base critical sections does not change noticeably among different excitation patterns. Taking into account

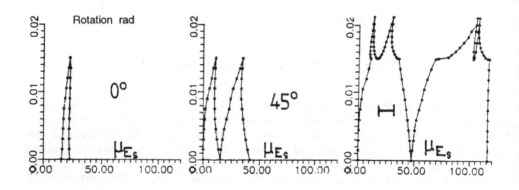

Figure 7. Strain energy ductility at the corner reinforcement

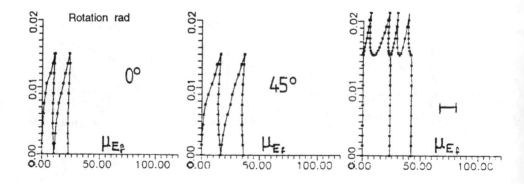

Figure 8. Flexural energy ductility at the base critical region

the strong relation between strain energy and absolute strain demand at the reinforcement (and hence overall damage of this resisting element), it can be concluded that generalized spatial excitation causes increased damage locally at the critical region, as also observed in the experiments, compared to planar behavior.

CONCLUSIONS

A square cantilever column is analyzed under different axial loads and deformation patterns in uniaxial and biaxial monotonic and cyclic response, keeping, for comparison, the imposed drift magnitude constant. The use of a refined finite element model allows for the estimation of flexurally induced damage along and across the member. Certain aspects regarding the damage characterization of different excitations for this column are demonstrated and are summarized as follows:

For the column considered, the assumption of a circular interaction is adequate for strength but not for onset of yield deformation, when this is expressed in terms of curvature in the base critical region.

As a result, vectorial deformation indices in use, normalized by the actual yield deformation prevailing under oblique excitation, indicate higher ductility demands, consistent with observation. Projected uniaxial ductilities are unable to depict such behavior.

The total flexural energy incurred in the system is independent of the angle of incidence. However, the distribution of this energy in the form of strain energy among the reinforcement depends strongly on Θ, pattern of loading and axial load. Generally, strain energy ductility correlates well with demanded reinforcement strains.

The biaxial response predictions of absorbed strain energy at the section level (E_s), unlike global response indices, corroborate better with observations of increased local damage compared to uniaxial response.

REFERENCES

1. Akiyama H., Earthquake-Resistant Limit State Design for Buildings, University of Tokyo Press, Tokyo, 1985.

2. Chung Y.S., Meyer C. and Shinozuka M., Modeling of Concrete Damage, ACI Structural Journal, 86, No. 3, May-June 1989, pp. 259-71.

3. Fajfar P. Vidic T. and Fischinger M., Seismic Demand in Medium-and Long-Period Structures, Earthq.. Eng. and Str. Dynamics, 18, 1989, pp.1133-44.

4. Low S.S. and Moehle J.P., Experimental Study of Reinforced Concrete Columns Subjected to Multi-Axial Cyclic Loading. Report EERC 87-14, University of California, Berkeley, 1987.

5. Mahin S. and Bertero V., An Evaluation of Inelastic Seismic Design Spectra, Jrnl. of the Str. Div., ASCE, 107, ST9. Sept. 1981, pp. 1777-95.

6. Mahin S. and Lin J.P., Construction of Inelastic Response Spectra for SDOF Systems, Report EERC 83-17, University of California, Berkeley, 1983.

7. Menegotto M. and Pinto, P., Slender RC Compressed Members in Biaxial Bending, Jrnl. of the Str. Div., ASCE, 103, ST3, March 1977, pp. 587-605.

8. Oliva M., Shaking Table Testing of a Reinforced Concrete Frame with Biaxial Response, Report EERC 80-28, Univ. of California, Berkeley, 1980.

9. Otani S. and Cheung V.W., Behavior of Reinforced Concrete Columns Under Biaxial Lateral Load Reversals, (ii) Test Without Axial Loads, Publ. 81-02. Univ. of Toronto, Toronto, Feb. 1982.

10. Park Y.J. and Ang A. H-S., Mechanistic Seismic Damage Model for Reinforced Concrete, Jrnl. of the Str. Div., ASCE, 111, ST4, April 1985, pp. 722-39.

11. Park Y.J., Wen Y.K. and Ang A.H-S., Random Vibration of Hysteretic Systems Under Bi-Directional Ground Motion, Earthq. Eng. and Str. Dynamics, 14, 1986, pp. 543-57.

12. Shahrooz B. and Moehle J., Experimental Study of Seismic Response of RC Setback Buildings, Report EERC 87-16, Univ. of California, Berkeley, 1987.

13. Uang C.M. and Bertero V., Use of Energy as a Design Criterion in Earthquake-Resistant Design, Report EERC 88-18, Univ. of California, Berkeley, 1988.

14. Zeris C.A., Three Dimensional Nonlinear Response of Reinforced Concrete Buildings. PhD thesis, Dept. of Civil Engineering, University of California, Berkeley, Dec. 1986.

15. Zeris C.A., An Investigation of the Nonlinear Biaxial Response of Simple Structural Systems, Proc. 4th U.S. Natl. Conf. of Earthq. Eng., 2, Palm Springs, May 1990, pp. 985-94.

16. Zeris C.A. and Mahin S.A., Behavior of Reinforced Concrete Structures Subjected to Biaxial Excitation, Jrnl. of the Str. Div., ASCE, 117, ST9, Sept. 1991.

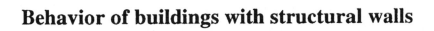

Behavior of buildings with structural walls

DESIGN OF R/C STRUCTURAL WALLS: BALANCING TOUGHNESS AND STIFFNESS

Sharon L. Wood
Department of Civil Engineering
University of Illinois
205 North Mathews Ave., Urbana, Illinois 61801, U.S.A.

ABSTRACT

The 1985 Chile earthquake provided a rare opportunity to study the seismic response of reinforced concrete structural wall buildings. Most of the reinforced concrete residential buildings relied on structural walls to resist vertical and lateral loads. However, the walls did not have the reinforcement details required in current U.S. codes to ensure ductile response. A survey of damage following the earthquake indicated that most of these buildings sustained no structural damage. Analyses of the buildings indicated that the structural walls provided sufficient stiffness to limit the earthquake damage. Relationships between the amount of wall area provided in a building and the expected displacement during an earthquake are discussed.

INTRODUCTION

For design of reinforced concrete walls in seismic zones, current practices in the U.S. and Chile represent opposite extremes. In the U.S., structural walls are most often used in frame–wall systems. Wall area is typically kept to a minimum to maintain architectural flexibility. Building codes [1, 2] require heavily confined boundary elements to ensure satisfactory performance of the walls during earthquakes (Fig. 1(a)). In contrast, structural walls are frequently used in Chilean construction to resist both lateral and vertical loads [3]. The amount of wall area is large, typically between 2 and 4% of the floor area in each direction. Reinforcement details in the Chilean walls are modest by U.S. standards (Fig. 1(b)).

If current U.S. design provisions [2] are used to evaluate representative Chilean buildings, the lack of confinement reinforcement in the walls would imply vulnerability to structural damage during a design–intensity earthquake. Walls in Chilean buildings do not have sufficient toughness to maintain their load–carrying capacity during large–amplitude displacement cycles. However, the excellent performance of moderate–rise buildings in Viña del Mar, Chile during the 1985 earthquake provides convincing evidence that toughness is not required in structural walls if the buildings have sufficient stiffness to limit the displacement response.

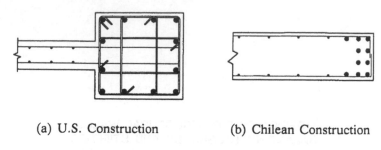

(a) U.S. Construction   (b) Chilean Construction

Figure 1. Representative Wall Reinforcement Details

A method is presented in this paper to interpret the displacement response of structural wall buildings subjected to earthquake loading. The expected displacement demand is then used to establish toughness and stiffness limits for satisfactory wall response.

BUILDING INVENTORY IN VIÑA DEL MAR

At the time of the 1985 Chile earthquake (M_s = 7.8), more than 165 reinforced concrete residential buildings, ranging in height from 6 to 23 stories, were located in Viña del Mar [3]. Structural walls were used to resist lateral and vertical loads in 98% of these buildings. A representative floor plan is shown in Fig. 2. The locations of the walls tended to follow the architectural layout of the buildings; structural walls were often used as the partitions between apartments. As a result of the large area of structural walls, the Chilean buildings tend to be very stiff. Measured periods are on the order of one half those measured in shear wall buildings of comparable height in the U.S. [4].

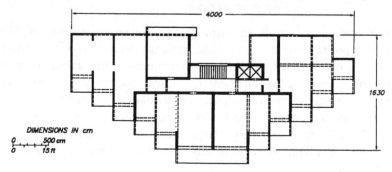

Figure 2. Floor Plan of Representative Building in Viña del Mar

Viña del Mar is located 80 km from the epicenter of the 1985 earthquake. Peak accelerations of 0.36g in the S20W direction and 0.23g in the N70W direction were recorded in the downtown area. Both components included approximately 40 seconds of motion in which the horizontal accelerations exceeded 0.1g.

Structural damage in Viña del Mar was light (Fig. 3) [3]. Five buildings sustained severe damage, eight experienced moderate structural damage, and light structural damage was observed in 21 buildings. One hundred thirty–one buildings, nearly 80% of the inventory survived the earthquake with no structural damage.

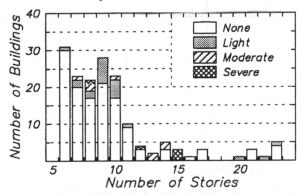

Figure 3. Observed Structural Damage in Viña del Mar [3]

A GENERALIZED INTERPRETATION OF STRUCTURAL RESPONSE

The satisfactory performance of buildings in Viña del Mar during the 1985 earthquake prompted an investigation of the required balance between toughness and stiffness in reinforced concrete walls. In the following section, a model is developed for interpreting the structural response of structural wall buildings subjected to strong ground motion. Displacement is used as an index of structural damage throughout this discussion.

A building in which structural walls represent the primary source of lateral stiffness may be modelled by the idealized structure shown in Fig. 4 [5]. The structural properties of all walls are represented by a single cantilever wall. The fundamental period of a cantilever wall with uniform distributions of mass and stiffness along its length is:

$$T = \frac{2\pi}{3.52} \sqrt{\frac{\mu H^4}{EI}} \tag{1}$$

where T is the fundamental period in seconds, H is the height of the wall, μ is the mass per unit length, E is Young's modulus, and I is the moment of inertia of the wall cross section.

Assuming a rectangular wall cross–section, that the self–weight of the wall is small compared with the floor dead load, and that floor loads are distributed uniformly over the tributary area, A_t, Eq. 1 may be rewritten as:

$$T = 6.18 * \frac{H}{D} * N * \sqrt{\frac{wh}{Epg}} \tag{2}$$

where D is the length of the wall, t is the wall thickness, N is the number of stories, w is the floor load per unit area, h is the typical story height, p is the ratio of wall area in one direction to the tributary floor area, and g is the gravitational constant.

Based on this formulation, fundamental period of structural wall buildings in Chile may be expressed in terms of three nondimensional parameters: the slenderness ratio of

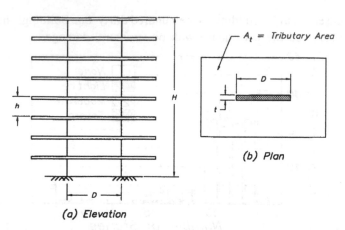

(a) Elevation

Figure 4. Conceptual Model of Structural Wall Building

the primary walls, H/D; the number of stories, N; and the ratio of wall area to floor area, p. Within the Viña del Mar inventory, the floor loads, story heights, and concrete strengths did not vary appreciably. Therefore, representative values of $w = 1000$ kg/m^2 (200 lb/ft^2), $h = 2.65$ m (104 $in.$), and $f'_c = 250$ kg/cm^2 (3500 psi) were used for all calculations. Young's modulus was taken to be 237,000 kg/cm^2 (3400 ksi) based on the relationship between concrete strength and stiffness given in ACI 318 [1].

Analytical [6] and experimental [7] studies have shown that the inelastic displacement response of a reinforced concrete system may be bounded by the displacement calculated using a linear response spectrum with a damping factor of 0.02 and a linear model of the structural system with an effective stiffness equal to one–half the initial, gross–section stiffness. The effective period, T_e, of the idealized wall structure may be written as:

$$T_e = \sqrt{2} * T = 8.74 * \frac{H}{D} * N * \sqrt{\frac{wh}{Epg}} \qquad (3)$$

Using this representation of effective period and a linear response spectrum, the sensitivity of the displacement response of structural wall buildings to wall slenderness, H/D, and wall area, p, may be evaluated.

Displacement Demand in Structural Wall Buildings

The displacement demand in a structural wall building during a severe earthquake in the western U.S. was evaluated using the response spectrum proposed by Shibata and Sozen [8]. The spectral acceleration for a damping factor of 0.02 is given by:

$$S_a = 1.5\frac{A_{max}}{T} \leq 3.75A_{max} \qquad (4)$$

where A_{max} is the effective peak base acceleration. The corresponding spectral displacement is:

$$S_d = 3.75 \frac{A_{max} T^2}{(2\pi)^2} \qquad T \le 0.4 \text{ sec} \qquad (5a)$$

$$S_d = 1.5 \frac{A_{max} T}{(2\pi)^2} \qquad T \ge 0.4 \text{ sec} \qquad (5b)$$

To simplify calculations, the spectral displacement was assumed to vary linearly with period for the entire range of periods considered (Eq. (5b)).

As shown in Fig. 5, the linear variation of spectral displacement with period is conservative in the short period region. The displacement spectrum corresponding to the north–south component of the 1940 El Centro earthquake is shown for comparison. The acceleration record was normalized such that the maximum acceleration was 0.5g.

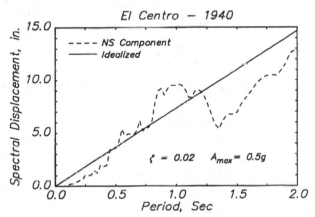

Figure 5. Linear Response Spectrum for Strong Ground Motion in Western U.S.

Building displacement response may be calculated using the displacement spectrum shown in Fig. 5. An amplification factor of 1.5 was used to convert from spectral displacement to displacement at the roof of the buildings. The maximum displacement was interpreted in terms of the mean drift ratio, the ratio of roof displacement to building height. Using the effective period given in Eq. (3) and an effective peak base acceleration of 0.5g, the mean drift ratio, MDR, may be expressed as:

$$MDR = \frac{1}{4} * \frac{H}{D} * \sqrt{\frac{wg}{Eph}} \qquad (6)$$

The mean drift ratio is plotted as a function of wall area in Fig. 6 for various slenderness ratios. The calculated displacement demand is sensitive to the amount of wall area. In the range of typical U.S. construction (wall areas between 0.5 and 1.0% of the floor area), the displacement demand exceeded 1% drift for all walls with slenderness ratios greater than 2. The calculated displacement demand decreased rapidly for wall areas between 1 and 2% of the floor area. The sensitivity of displacement demand to wall area decreased between 2 and 4% wall area, and drift was nearly independent of wall area for buildings with more than 4% wall area. Within the range of wall areas found in Chilean construction (2 to 4% of the floor area), displacement demand was less than 1% of the height of the building for walls with slenderness ratios up to 5.

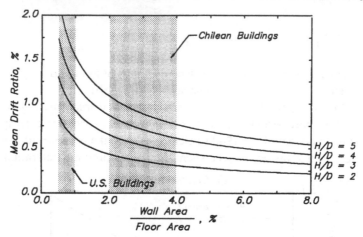

Figure 6. Calculated Displacement Demand in Structural Wall Buildings

Displacement Capacity of Slender Walls

The analyses described in the previous section indicated that the maximum displacement demand in a structural wall building during a strong earthquake may range from 1% of the height in Chilean buildings to 2% of the height in U.S. buildings. Therefore, the displacement capacity of structural walls becomes an important parameter for evaluating the seismic response of structural wall buildings. Data from twenty–seven isolated walls with slenderness ratios between 2 and 3 [9, 10, 11, 12, 13, 14, 15] were considered to evaluate the relationship between wall toughness and reinforcement details. All walls were fixed at the base and loaded laterally with a single force at the top. Walls with rectangular, barbell, flanged, and C–shaped cross sections were considered. Four walls had openings.

The displacement capacity was defined as the maximum displacement that the wall could sustain for two complete loading cycles with less than a 20% decrease in lateral strength. Nearly half the wall were constructed with less transverse steel in the boundary elements that required in current U.S. Codes [1, 2] for confinement. The confinement index is defined as the ratio of the area of transverse steel provided in the boundary elements to the area required to satisfy the special detailing provisions of the UBC [2].

Measured displacement capacity is plotted as a function of the confinement index in Fig. 7. The displacement capacity of all walls exceeded 1% of their height. Many factors influence the displacement capacity in addition to the confinement index, including the mode of failure and the amount of axial load. However the data in Fig. 7 suggest that structural walls of these typical configurations can sustain displacements on the order of 1% of their height without special transverse reinforcement.

The laboratory data confirm the experience of the Viña del Mar inventory: if displacements are controlled, special reinforcement details are not required to ensure sat-

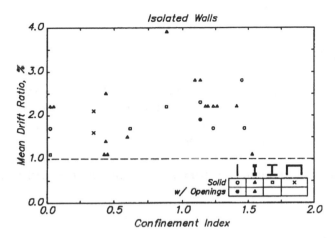

Figure 7. Displacement Capacity of Laboratory Specimens

isfactory structural performance. The displacements in Chilean buildings are controlled by providing 2 to 4% wall area in each direction of the building. The Chilean buildings are not likely to perform well if displacements exceed 2% of the height, but buildings with that little wall area are not used in Chile.

The laboratory data also indicate that providing confinement of the boundary elements may not be sufficient to ensure that structural walls can resist displacements on the order of 2% of their height. Shear failures were observed in three of the four walls with a confinement index greater than 1.0 that failed at displacement levels between 1 and 2% drift. The toughness required in structural walls used in U.S. buildings may be limited by web crushing failures, in addition to instability of the boundary elements.

STIFFNESS AS A DESIGN CRITERION FOR STRUCTURAL WALLS

Nothing in the current seismic building codes [2] precludes engineers from designing and constructing buildings with large amounts of wall area in the U.S. The buildings are classified as concrete bearing wall systems, and lateral forces are calculated using an R_w factor of 6 [2]. However, moderate-rise buildings with 2 to 4% wall area are not common in the U.S. The primary reason that these buildings are not constructed is the expense of providing boundary elements at the end of each wall. The requirements for boundary elements are based on construction of frame-wall systems where much larger displacement capacities are needed. There are no provisions in the current codes to take advantage of the stiffness provided by a large number of walls to limit the displacements during an earthquake.

An indication of the type of buildings for which boundary elements would be required by UBC [2] provisions may be obtained by considering the idealized structure shown in Fig. 4. The design base shear for this building would be:

$$V_b = \frac{ZIC}{R_w} W \tag{7}$$

where

$$C = \frac{1.25S}{T^{2/3}} \leq 2.75 \tag{8}$$

In this example, the zone factor Z is assumed to be 0.4, the occupancy factor I is assumed to be 1.0, R_w is 6, the soil factor S is assumed to be 1.2, and the period T is given by Eq. (1). The design base shear, is therefore, a function of wall slenderness, H/D, wall area, p, and the number of stories in the building, N.

If the design lateral forces are assumed to increase linearly with height above the base. Therefore, the overturning moment that the wall must be designed to resist is given by:

$$M_b = V_b * \frac{2}{3}H \tag{9}$$

The UBC [2] requirements state that boundary elements are required in walls for which the factored compressive stress at the extreme fiber exceeds $0.2f'_c$ under design loads. Assuming a critical load combination of 1.4 times the dead load plus 1.7 times the earthquake load, the maximum factored compressive stress, σ, is

$$\sigma = 1.7\frac{M_b c}{I} + 1.4\frac{W}{A_w} \tag{10}$$

where $c = D/2$, $I = tD^3/12$, the axial load $W = NA_t w$, and the area of the wall $A_w = Dt$. Compressive stress, is therefore, a function of three parameters: H/D, p, and N.

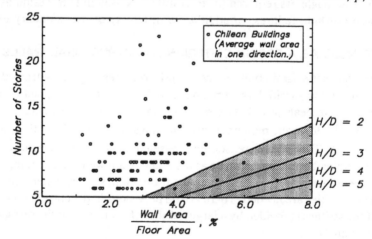

Figure 8. Uniform Building Code [2] Requirements for Boundary Elements in Structural Walls

The population of buildings in Viña del Mar is represented by circles in Fig. 8. The shaded region represents the combination of stories and wall area for which the factored compressive stress is less than $0.2f'_c$ for a wall with a slenderness ratio of 2. Lines are also shown for slenderness ratios up to 5. Only low–rise buildings with large amounts of wall area would be exempt from the UBC [2] boundary element provisions.

Boundary elements would have been required in virtually all the buildings in Viña del Mar if the buildings had been designed in the U.S. The UBC [2] requirements are clearly too stringent for buildings with large amounts of wall area. The demonstrated satisfactory performance of the Chilean buildings and results of laboratory tests indicate that confined boundary elements are not required if drift is controlled. Using the Chilean approach of providing 2 to 4% wall area in each direction has been shown to control displacements for a strong earthquake in the western U.S. (Fig. 8).

CONCLUSION

The performance of reinforced concrete buildings during the 1985 Chile earthquake and the associated studies of generalized wall response have demonstrated that building stiffness can eliminate the need for toughness in structural walls. The option of using drift as a criterion for determining the need for confined boundary elements in walls deserves consideration.

ACKNOWLEDGMENTS

The work described in this paper was performed as part of a cooperative research project sponsored by the National Science Foundation to investigate the performance of reinforced concrete buildings during the 1985 Chile earthquake. The comments of researchers associated with this project (J.P. Moehle, University of California at Berkeley; R. Riddell, Universidad Católica de Chile; M.A. Sozen, University of Illinois at Urbana –Champaign; and J.K. Wight, University of Michigan) are greatly appreciated.

REFERENCES

1. American Concrete Institute Committee 318, "Building Code Requirements for Reinforced Concrete (ACI 318–89)." American Concrete Institute, Detroit, 1989.

2. *Uniform Building Code*, International Conference of Building Officials, Whittier, California, 1988.

3. Riddell, R.; Wood, S.L.; and De La Llera, J.C., "The 1985 Chile Earthquake: Structural Characteristics and Damage Statistics for the Building Inventory in Viña del Mar." Civil Engineering Studies, *Structural Research Series No. 534*, University of Illinois, Urbana, 1987.

4. Wood, S.L. and J.P. Moehle, "Performance of R/C Shear Walls Buildings During the 1985 Chile Earthquake," 1988 Earthquake Engineering Research Institute Annual Meeting Seminar, Mesa, Arizona, February 1988.

5. Sozen, M.A., "Earthquake Response of Buildings with Robust Walls." *Proceedings*, Fifth Chilean Conference on Seismology and Earthquake Engineering, Santiago, 1989.

6. Shimazaki, K. and Sozen, M.A., "Seismic Drift of Reinforced Concrete Structures." *Research Reports*, Hazama–Gumi Ltd., Tokyo, 1984, pp. 145–166. (in Japanese).

7. Bonacci, J.F., "Experiments to Study Seismic Drift of Reinforced Concrete Structures." Ph.D. Dissertation submitted to the Graduate College of the University of Illinois, Urbana, 1988.

8. Shibata, A. and Sozen, M.A., "Substitute Structure Method for Seismic Design in Reinforced Concrete." *Journal of the Structural Division*, American Society of Civil Engineers, Vol. 102, No. ST1, 1976, pp. 1–18.

9. Ali, A. and Wight, J.K., "Reinforced Concrete Structural Walls with Staggered Opening Configurations under Reversed Cyclic Loading." *Report No. UMCE 90–05*, Department of Civil Engineering, University of Michigan, Ann Arbor, 1990.

10. Lackner, W.R. and Wood, S.L., "Cyclic Behavior of Reinforced Concrete Core Walls." Civil Engineering Studies, *Structural Research Series*, University of Illinois, Urbana, 1991.

11. Morgan, B.J.; Hiraishi, H.; and Corley, W.G., "U.S.–Japan Quasi-Static Test of Isolated Wall Planar Reinforced Concrete Structure." Report to the National Science Foundation, Construction Technology Laboratories, Portland Cement Association, Skokie, Illinois, 1986.

12. Oesterle, R.G., "Inelastic Analysis for In–Plane Strength of Reinforced Concrete Shear Walls." Ph.D. Dissertation submitted to the Graduate School of Northwestern University, Evanston, Illinois, 1986.

13. Oesterle, R.G.; Fiorato, A.E.; Johal, L.S.; Carpenter, J.E.; Russell, H.G.; and Corley, W.G., "Earthquake Resistant Structural Walls – Tests of Isolated Walls." Report to the National Science Foundation, Construction Technology Laboratories, Portland Cement Association, Skokie, Illinois, 1976.

14. Oesterle, R.G.; Aristizabal–Ochoa, J.D.; Fiorato, A.E.; Russell, H.G.; and Corley, W.G., "Earthquake Resistant Structural Walls – Tests of Isolated Walls – Phase II." Report to the National Science Foundation, Construction Technology Laboratories, Portland Cement Association, Skokie, Illinois, 1979.

15. Shiu, K.N.; Daniel, J.I.; Aristizabal–Ochoa, J.D.; Fiorato, A.E.; and Corley, W.G., "Earthquake Resistant Structural Walls – Tests of Walls with and without Openings." Report to the National Science Foundation, Construction Technology Laboratories, Portland Cement Association, Skokie, Illinois, 1981.

ADVANCES IN THE DESIGN FOR SHEAR OF RC STRUCTURAL WALLS UNDER SEISMIC LOADING

EINAR KEINTZEL
Institut für Massivbau und Baustofftechnologie,
Universität Karlsruhe, W 7500 Karlsruhe, Germany

ABSTRACT

The paper deals with the determination of design shear forces in RC structural walls, yielding under seismic loading. At first a survey of different approches to the evaluation of seismic design shear forces for RC walls is given. Then extensive parametric studies on this subject, by means of nonlinear time history calculations, are presented. From an attempt to explain the results of these calculations the concept of modal limit forces is developed. Using this concept, an approximate representation of shear forces in yielding cantilever shear walls by response spectrum analysis is proposed, with modal participation factors different from those in the elastic range. A simplified formula is also given, to be applied in the case of a seismic analysis, considering only the fundamental mode. Finally the subject of the paper is considered in the more general framework of some analogous problems in cantilever structures.

INTRODUCTION

Experimental investigations of RC members under cyclic deformations in the inelastic range have shown that substantial strength degradations can be avoided only if the inelastic behaviour is controlled by flexure, and not by shear. Thus for the seismic design of RC structural members it is recommended frequently to consider design shear forces corresponding to their flexural strengths, respectively, as an upper limit, to the elastic behaviour of the structure under seismic loads.

Whereas in framed structures the shear forces corresponding to the flexural strength of beams and columns can be evaluated simply from the yielding moments of their end sections, in cantilever wall structures a more complicated situation is encountered. So in the case of a n-storied shear wall structure, reduced as in Fig. 1a to a cantilever, the shear force at flexural strength depends as well on the yielding moment of the base section as on the position of the centre of seismic loads.

First investigations of seismic shear forces in RC shear wall structures have been performed in [1]. They have led for a given yielding moment to shear forces, higher than those which would correspond to the distribution of seismic loads in the elastic range, pointing to an increased importance of higher vibrational modes in case of inelastic behaviour. A series of further investigations [2-5] have led to similar conclusions. So the results of nonlinear time history analyses for RC structural walls in [2], [4], [5] have shown that the shear in the critical region at the base of walls fluctuates more rapidly with time than either moment or rotation, being more sensitive to higher mode response. In [1-5] shear magnification factors have been proposed, expressing the shifting of the centre of seismic loads at yield in a position lower than that corresponding to elastic behaviour, due to the increased importance of higher modes.

Magnification factors for shear forces based on [1] have been introduced in several national codes as well as in the CEB code [6]. So the shear force Q at the base of a RC shear wall is determined in the CEB code by the relations (written slightly transformed)

$$Q \leq \omega\, Q_d\, M_y/M_d \tag{1}$$

$$Q \leq Q_o \tag{2}$$

with
Q_d, M_d – design values of the seismic shear force and the bending moment, calculated by equivalent static analysis, considering only the fundamental vibrational mode,
ω – magnification factor, taking into account the dynamic inelastic behaviour of the wall, with
$\omega = 0.9 + 0.1\, n$ for $n \leq 5$ and
$\omega = 1.2 + 0.04\, n \leq 1.8$ for $n > 5$,   (3)
M_y – yielding moment,
Q_o – seismic shear force in the elastic range.

Subsequent investigations have shown a series of inadequacies in the approximate relations (1) to (3). So it was shown in [7] that the shear forces do not increase proportionally with the yielding moment of the shear walls, as supposed in Eq. (1), but more slowly, and that for $n \geq 5$ the magnification factor does not depend in fact on n, but, by means of n, on the fundamental period. Similarly it was shown in [3], [5], that the dynamic magnification of shear forces is larger for higher seismic input level, respectively for higher ductility factors.

In this paper, in order to establish an improved representation of the mentioned shear magnification, an attempt is made to develop approximate relations for the determination of seismic design shear forces in RC yielding wall structures by response spectrum analysis, corresponding as well as possible to the results of extensive parametric investigations [8], [9], the assumptions and results of which are presented in the following.

PARAMETRIC INVESTIGATIONS

Structural Model
For the mentioned investigations shear wall structures have been reduced, as shown in Fig. 1a, to cantilevers with n = 2, 3, 4 and 5 lumped story masses m at equal distances on the height h. The yielding moment of a cantilever is given by the relation

$$M_y = c M_1 g / S_{ad}(T_1) \tag{3}$$

with

c   – yielding level factor;

$S_{ad}(T_1)$ – design value of the acceleration response spectrum for the fundamental vibrational period T_1 of the structure;

M_1 – overturning moment at the base of the structure, due to the design seismic load, corresponding to the fundamental vibrational mode;

g   – acceleration of gravity.

For a system with n = 1, having the yielding force F_y and the yielding moment M_y = hF_y, Eq. (1) turns into F_y = cmg. For each number of story masses 4 different yielding levels (3 values of c in Fig. 1b and for comparison elastic behaviour), 4 values of T_1 (0.2 s, 0.4 s, 0.8 s and 1.6 s) and 5 values of the index α = EI/GAh^2 (0.00, 0.25, 0.50, 0.75 and 1.00) with EI, GA flexural resp. shear stiffness are considered. The yielding level 2 corresponds to the design forces of the German Seismic Code DIN 4149 [10] for the MSK intensity I = 8. For response spectrum calculations the damping ratio D = 0.05 is admitted, and for time history calculations Rayleigh damping with the first two damping ratios D_1 = D_2 = 0,05. Conservatively only flexural yielding at the base, and not also shear yielding, is considered. For the case of elastic behaviour under seismic loading the base shear, belonging to the vibrational mode i with the period T_i, may be expressed as

$$Q_{io} = \varepsilon_i \, n \, m \, S_a(T_i), \tag{4}$$

where ε_i is an effective mass factor and $S_a(T_i)$ the elastic acceleration

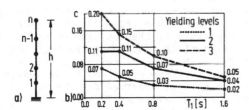

Fig. 1 Structural characteristics.
    a) Structural systems,
    b) yielding levels

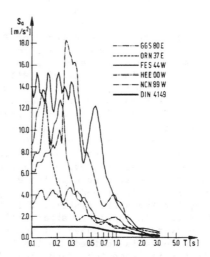

Fig. 2 Response spectra for 5% damping

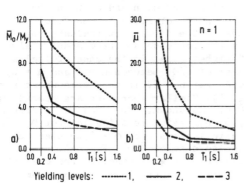

Yielding levels: ········ 1,   ——— 2,   ——— 3

Fig. 3 Yielding of the system under seismic excitation.
    a) Ratios $\overline{M}_o/M_y$,
    b) displacement ductility factors μ

response spectrum value.

Earthquake Excitation and Inelastic Response Behaviour

The 320 structural variants, defined, as shown before, by n, c, T_1 and α, have been analysed under the acceleration time histories of a set of 10 strong motion records, belonging to 5 real earthquakes: San Francisco Golden Gate Park 1957. 3.22 (GG, I = 7, M = 5.3), Helena (Montana) 1935. 10.31 (HE, I = 8, M = 6.0), Hollister (Northern California) 1949. 3.9 (NC, I = 7, M = 5.2), Ferndale (California) 1975. 6.7 (FE, I = 7, M = 5.7) and Oroville (California) 1975. 8.1 (OR, I = 7, M = 5.8). In order to facilitate the observation of the influences of higher modes, the earthquakes have been chosen in the range of relative low magnitudes, with predominant higher frequencies of the ground shaking. In Fig. 2 acceleration response spectra of the stronger components of each earthquake time history, scaled for the MSK intensity I = 8, are represented. In Fig. 3 the strength and ductility of the structural models of Fig. 1 are related to the earthquake excitation, defined as before. Fig. 3a shows mean values of the flexural moment at the base in the elastic range M_o, related to M_y, and Fig. 3b mean values of displacement ductility factors μ for SDF systems (n = 1).

Results

Examples of results, represented as ratio between the mean value $\bar{Q}$ of the base shear, computed by nonlinear time history analysis, and the design value Q_1 of the German Seismic Code DIN 4149 [10] (yielding level 2), corresponding to the fundamental vibrational mode of a structure with $\alpha = 0$, are represented in Fig. 4 to 6. The similarity between the shapes of the curves, belonging to inelastic and to elastic behaviour, is noticed. The peculiarities of the curves corresponding to elastic behaviour can be explained by means of the characteristics of higher modes in the modal analysis, shown in Fig. 7. So the increase of $\bar{Q}/Q_1$ with n (Fig. 5), important especially for $\alpha = 0$ and $T_1 = 1.6$ s, is explained by the increase of $\varepsilon_2/\varepsilon_1$ (Fig. 7a). Further the increase of T_2/T_1 with α (Fig. 7b) leads for high values of T_1 to a decrease of Q_{2o}, because T_2 reaches in the region of the descending branch of the acceleration response spectrum curve and ε_2 decreases also. Thus the increase of $\bar{Q}/Q_1$ with T_1 due to Q_{2o}, observed for $\alpha = 0$, is not found for $\alpha = 0,25$ (Fig. 4). However for low values of T_1 the period T_2 remains in the region of the ascending branch of $S_a(T)$ and Q_{2o} increases with α. Thus the increase of α leads for low values of T_1 to an increase and for high values of T_1 to a decrease of $\bar{Q}/Q_1$ (Fig. 6).

A comparison of the values $\bar{Q}/Q_1$ in Fig. 4 to 6, obtained for elastic and for inelastic systems, shows that the influence of higher modes on the shear force of elastic systems is found qualitatively also in the case of inelastic systems. Even niceties as the influences of n and of α on the parameters of higher modes (T_2/T_1, $\varepsilon_2/\varepsilon_1$) can be observed in the diagrams of $\bar{Q}/Q_1$ for inelastic systems. This leads to the idea to approximate the shear forces in inelastic shear walls by the response spectrum modal analysis. However, an increased influence of higher modes in the case of inelastic structures is noticed. So the increase of $\bar{Q}/Q_1$ between n = 1 and n = 2, due to the intervention of the second mode for n = 2, is more pronounced for inelastic than for elastic systems (Fig. 5). Similar phenomena are stated in connection with the influence of α on $\bar{Q}/Q_1$. As shown, $\bar{Q}/Q_1$ does not increase with T_1 for elastic systems with $\alpha = 0,25$, Q_{2o} being relative low for $\alpha > 0$, (Fig. 4 and Fig. 7b). However for inelastic systems with $\alpha = 0,25$ even the low values of Q_{2o} are able to produce an increase of $\bar{Q}/Q_1$ (Fig. 4). The consideration of the increased influence of higher modes in the case of inelastic structures implies some modifications in the response spectrum method.

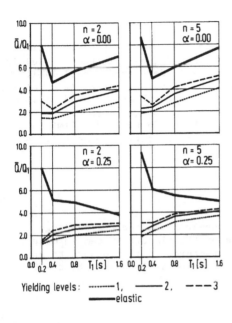

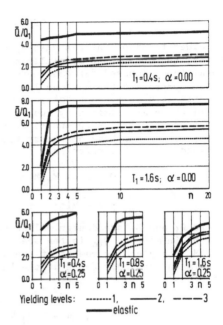

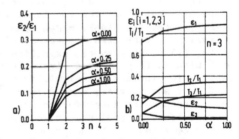

Fig. 4 Shear forces as a function of T_1

Fig. 5 Shear forces as a function of n

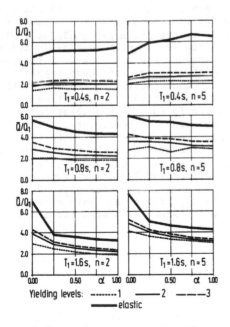

Fig. 6 Shear forces as a function of α

Fig. 7 Characteristics of higher modes for elastic systems.
a) Effective mass factors as a function of n,
b) effective mass factors and period ratios as a function of α

APPROXIMATE RELATIONS

In order to adapt the response spectrum modal analysis to yielding structures, the concept of modal limit forces (modal forces, limited by yielding
conditions, imposed for each mode separately) is developed. Modal limit
forces represent the maximum limit values that internal forces in an elastoplastic structure can attain, if only one certain vibrational mode i is excited. They are limited by two physical features: by the strength of the
structure and by the intensity of the excitation. So the modal limit shear
force Q_{iy} at the base of the system in Fig. 1a, belonging to mode i, is
defined by the two relations

$$Q_{iy} \leq Q_i \, M_y/M_i \quad \text{and} \tag{5}$$

$$Q_{iy} \leq Q_{io}, \tag{6}$$

where Q_i and M_i are design values of the shear force and the bending
moment in mode i, Q_{io} is the shear force of the elastic system in mode i,
and M_y the yielding moment of the wall structure. It is proposed to approximate the shear force at the base of the inelastic system by the SRSS
relation

$$Q = \gamma \, Q^* = \gamma \sqrt{\sum_{i=1}^{n} Q_{iy}^2}, \tag{7}$$

where γ is a correction factor. Q^* corresponds to a response spectrum modal
analysis, in which the modal forces are considered separately until yielding.

Eq. (5) and (6) may be considered as generalization of Eq. (1) and (2).
As usually the condition (5) is decisive for the fundamental mode, but the
condition (6) for higher modes, the form of the relation (7) explains qualitatively the increased importance of higher modes, taken into consideration
in Eq. (1) roughly by the magnification factor ω, the relative weak increase
of $\overline{Q}$ with M_y (Fig. 1b and Fig. 4 to 6) and the increase of shear magnification with increasing seismic input level. Quantitative aspects are shown
in Fig. 8 and 9, where mean values $\overline{Q^*/Q}$ and coefficients of variation V of
the ratio between shear forces Q^*, calculated by Eq.(7), and shear forces
Q, calculated by nonlinear time history analysis, are represented. For elastic systems and $\gamma = 1$ the proposed procedure becomes identical with the
usual response spectrum modal analysis in connection with the SRSS formula.
For inelastic systems it leads to results of an exactitude comparable with
that obtained for elastic systems. Only for yielding level 1 systems, with
extremely high plastic deformations, unacceptable in practice (Fig. 3), the
exactitude is worse. In order to render the approximate analysis results for
inelastic systems at least as conservative as those calculated for elastic
systems, a correction factor

$$\gamma = \frac{\overline{(Q^*/Q)}_o}{\overline{Q^*/Q}} \, \frac{1 + V}{1 + V_o} \tag{8}$$

is introduced, where Q^*/Q and V correspond to inelastic and $(Q^*/Q)_o$ as well
as V_o to elastic systems. In Fig. 10 calculated values γ are represented together with the approximate expression

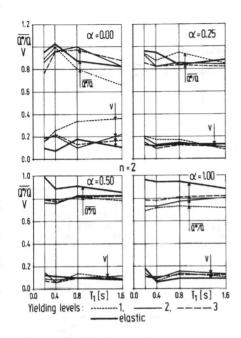 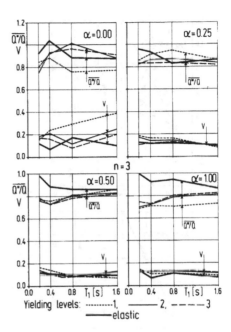

Fig. 8 Comparison response
spectrum versus time
history analysis for n=2

Fig. 9 Comparison response
spectrum versus time
history analysis for n=3

$$\gamma = 1 + 0.1 \ (M_o/M_y - 1) \geq 1. \tag{9}$$

In code design, according to usual assumptions, may be considered $\gamma = 1$.

For the case of simplified dynamic analysis, taking into account only the fundamental mode, a simplified relation is given, introducing the second mode implicitely. As on the one hand the simplified dynamic analysis is usually allowed to be applied only for systems with $T_1 \leq 2 \ T^*$ (Fig. 11), but on the other hand the second period T_2 amounts approximately $T_1/6$, for the simplified design spectrum of Fig. 11 is found always $S_{ad}(T_2) = \max S_{ad}$. Than the seismic shear force at the base can be approximated as $Q = \omega Q_I$, where

$$\omega = q\gamma \sqrt{(M_y/qM_I)^2 + 0.1 \ (\max S_{ad}/S_{ad}(T_1))^2} \leq q, \tag{10}$$

Q_I and M_I are design values of the shear force and the bending moment, according to the simplified dynamic analysis, and q is the behaviour factor, introduced in Eurocode 8 [11]. The factor 0.1 in the second term under the square root is determined in [9] as a conservative value, considering different approximate assumptions of the simplified dynamic analysis and different numbers of stories. The fact that q appears only in the first term under the square root denotes that only the contribution of the fundamental mode is considered reduced by yielding. Hence the increased importance of the second mode (represented by the second term) at yielding systems.

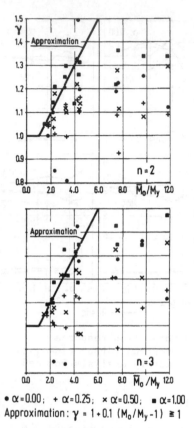

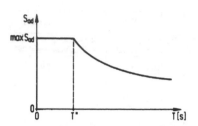

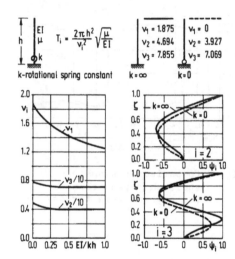

Fig. 11 Simplified design spectrum

• $\alpha = 0.00$; + $\alpha = 0.25$; × $\alpha = 0.50$; ■ $\alpha = 1.00$

Approximation: $\gamma = 1 + 0.1\,(M_o/M_y - 1) \geq 1$

Fig. 10 Correction factors γ
for n=2 and n=3

Fig. 12 Vibrational modes of elastic
cantilevers with a rotational
spring at the base

ANALOGOUS PROBLEMS

The mechanical meaning of the seismic shear magnification at yielding wall
structures, investigated in this paper, can be elucidated by examining it
in the context of analogous problems, appearing at cantilever structures.
In the dynamic behaviour of cantilever structures the fixing condition at
the base influences the fundamental vibrational mode in general strongly,
but the higher modes only slightly. In Fig. 12 this phenomenon is illustrated for cantilevers in the linear-elastic range. On the left the first three
vibrational periods T_i are given for cantilevers with a rotational spring at
the base. The coefficients v_i, met in the expression of T_i, are represented
in a diagram [12]. Whereas the value v_1, corresponding to the fundamental
mode, decreases strongly, if the parameter EI/kh increases, the decrease of
v_2 and v_3 is scarcely perceptible. On the right numerical values v_i and
modal shapes are represented for the limit values of the spring constant
$k = \infty$ (cantilever fixed at the base) and $k = 0$ (cantilever hinged at the
base). Whereas no proper fundamental mode can be attributed to the hinged
cantilever, its higher modal shapes do not differ very much from those of

the fixed cantilever. It can be shown by nonlinear calculations that the mentioned phenomenon, demonstrated above for the linear-elastic range, appears in some extent also in the nonlinear range and makes it possible to elaborate on this basis approximate solutions for different problems of nonlinear dynamics. As an example may be considered a simplified procedure for the evaluation of the base shear at storied structures with soil-structure interaction and foundation uplift under seismic excitation [13], [14]. In this procedure the effects of soil-structure interaction and foundation uplift are introduced in an approximate modal representation only for the fundamental mode and disregarded for higher modes.

The dynamic behaviour of yielding cantilever wall structures under seismic excitation, investigated in this paper, represents another example for the appearance of the mentioned phenomenon in the nonlinear range. In this case the fundamental mode is strongly influenced by yielding at the base and the shear force, belonging to it in the elastic range, is reduced according to Eq. (5) in order to be compatible with the flexural strength of the wall. On the other hand the higher modes practically are not influenced by yielding at the base; so the shear forces, belonging to them in the elastic range, are introduced unreduced in calculation according to Eq. (6). Detailed numerical investigations have led to the establishment of applicability conditions for this procedure, expressed by means of the concept of modal limit forces. According to this concept the shear force in the elastic range in mode i, considered in Eq. (6), can be introduced unreduced in calculation, if it is inferior to the shear force corresponding to yielding under the excitation only of mode i, given by Eq. (5); otherwise it is replaced by this shear force. As follows from Fig. 9 and 10, the proposed representation of shear forces in yielding RC cantilever structures leads in the range of practically available ductilities to results of acceptable exactitude.

CONCLUSIONS

An attempt is made to develop approximate relations for the determination of seismic design shear forces in RC yielding wall structures, corresponding as well as possible to the results of nonlinear time history calculations. A modal representation is proposed, in which the reduction of the elastic shear force by yielding is applied usually only in the fundamental mode. This is in accordance with the general observation that the higher vibrational modes at cantilever structures are influenced only slightly by the fixing condition at the base. The proposed relations have been introduced in the seismic code of the European Communities EUROCODE 8 [11].

REFERENCES

1. Blakeley, R.W.G., Cooney, R.C. and Megget, L.M., Seismic Shear Loading at Flexural Capacity in Cantilever Wall Structures. Bulletin of the New Zealand National Society of Earthquake Engineering, 1975, 8, pp. 278-90

2. Derecho, A.T., Iqbal, M., Ghosh, S.K., Fintel, M., Corley, W.G. and Scanlon, A., Structural Walls in Earthquake-Resistant Buildings - Dynamic Analysis of Isolated Structural Walls - Developement of Design Procedure - Design Force Levels, Final Report to the National Science Foundation, ASRA, under Grant No. ENV77-15333, Portland Cement Association, July 1981

3. Derecho, A.T. and Corley, W.G., Design Requirements for Structural
    Walls in Multistory Buildings. 8 WCEE, San Francisco 1989, Vol. 5,
    pp. 541-48

4. Kabeyasawa, T. and Ogata, K., Ultimate-State Design of R/C Wall-Frame
    Structures. Transactions of the Japan Concrete Institute, 1984, 6,
    pp.629-36

5. Aoyama, H., Earthquake Resistant Design of Reinforced Concrete Frame
    Buildings with "Flexural" Walls. Journal of the Faculty of
    Engineering, the University of Tokyo (B), 1987, 39, pp. 87-109

6. CEB Model Code for Seismic Design of Concrete Structures. Bulletin
    d'Information CEB No. 165. Lausanne, 1985

7. Keintzel, E., Ductility Requirements for Shear Wall Structures in
    Seismic Areas. 8 WCEE, San Francisco 1984, Vol. 4, pp. 671-77

8. Keintzel, E., Zur Querkraftbeanspruchung von Stahlbeton-Wandscheiben
    unter Erdbebenlasten. Beton- und Stahlbetonbau, 1988, 83, pp. 181-85,
    225-28

9. Keintzel, E., Vereinfachte Ermittlung der Querkraft aus Erdbebenlasten
    in Stahlbeton-Wandscheiben. Beton- und Stahlbetonbau, 1988, 83,
    pp. 324-26

10. DIN 4149, Teil 1: Bauten in deutschen Erdbebengebieten. Lastannahmen,
    Bemessung und Ausführung üblicher Hochbauten. Ausgabe April 1981

11. EUROCODE 8,Structures in Seismic Regions-Design. Part 1, General and
    building, May 1988 edition. Commission of the European Communities,
    1989

12. Keintzel, E., Die Berechnung der Schwingungen von Vielgeschoßbauten
    mit Hilfe kontinuierlicher Ersatzstrukturen. Die Bautechnik, 1967, 44,
    pp. 420-27

13. Yim, S.C-S. and Chopra A.K., Simplified Earthquake Analysis of
    Multistory Structures with Foundation Uplift. J. struct. eng. ASCE,
    1985, 111, pp. 2708-31

14. Chopra, A.K., Simplified Earthquake Analysis of Buildings. In Dynamics
    of Structures. Proceedings of the sessions at Structures Congress '87
    related to Dynamics of Structures, ed. M.J. Roesset, Orlando, Florida,
    1987, pp. 139-55

REQUIRED SHEAR STRENGTH OF EARTHQUAKE-RESISTANT REINFORCED CONCRETE SHEARWALLS

S. K. GHOSH
Portland Cement Association
Skokie, Illinois 60077, U.S.A.

ABSTRACT

A major unsolved aspect of the earthquake resistant design of buildings with reinforced concrete shearwalls is the evaluation of the shear force to be used in the design of the walls to avoid shear failure. This paper discusses dynamic response analysis to evaluate maximum shear forces in the walls, which will ultimately lead to suitable design recommendations.

INTRODUCTION

Shearwalls in multistory buildings are slender, and behave essentially as vertical cantilever beams. They tend to yield first at the base where the moment is the greatest. In earthquake-resistant design, to assure inelastic deformability, it becomes necessary to ensure that no brittle shear (diagonal tension) failure would develop prior to or simultaneously with the development of flexural hinging at the base of a wall.

Proper shear design for a structural element involves satisfaction of the inequality:

$$V_u \leq \phi V_n$$

(1)

where V_u is the required shear strength at a critical section, V_n is the nominal shear strength of that section, and ϕ is a strength reduction factor.

In the case of beams, the required shear strength is determined by assuming that moments of opposite sign corresponding to probable strength act at the joint faces and that the member is loaded with the tributary gravity load along the span. Similarly, for a column, the required shear strength is determined from the consideration of maximum developable moments, consistent with the axial force on the column, occurring at the column ends. In the case of shearwalls, a similar design condition is not readily established. This is because the magnitude of the shear at the base of a wall (or at any level above) is dependent not only on the maximum developable flexural strength at the section, but on the location of the resultant of the earthquake-induced horizontal forces that are distributed along the height of the wall.

The aim of the ongoing research described here is to establish design values of dynamic shears (required shear strengths) at the bases of concrete shearwalls in buildings subject to seismic ground motion.

PREVIOUS WORK

Aoyama, in an important review paper [1], has summarized significant Japanese work on the subject of discussion.

An analytical program was developed to analyze single-degree-of-freedom pseudo-dynamic response of the full-scale seven-story building tested as part of the U.S.-Japan Cooperative Earthquake Engineering Program Utilizing Large Scale Testing facilities (1981-82) [2]. The building, consisting of three parallel three-bay frames in the direction of testing, with the middle panel of the middle frame in-filled with a shearwall, was fixed to the rigid test floor of the laboratory, the lateral loads were distributed in an inverted triangular shape, and the amplitude was varied according to a single-degree-of-freedom pseudo-dynamic load control based on the prescribed earthquake input motion. The analysis proved sufficiently accurate in simulating response observed in the testing. Then seven-degrees-of-freedom inelastic dynamic response of the test structure was analyzed. It was found that, although the overall displacement response was quite similar to the single-degree-of-freedom response, or in other words fundamental mode response, the base shear was more susceptible to the higher modes. The fluctuation due to higher modes was largely carried by the wall, thus increasing the imposed shear force in the wall.

A practical evaluation of maximum dynamic shear in the wall was attempted. The dynamic base shear was decomposed into a part associated with the inverted triangular first mode and another part associated with the fluctuating higher modes. It was assumed that absolute accelerations for the second and higher modes could be expressed as products of the input acceleration and suitable response magnification factors. It was further assumed that the response magnification factor for the second mode had values between 1.5 and 2.0, typically 1.7, and that those for the higher modes had values close to unity. It was found that the maximum value of the dynamic base shear did not exceed the sum of the modal maxima:

$$V_{max} \leq V_{1max} + V_{Fmax} \qquad (2)$$

where V_{max} is the maximum dynamic base shear, V_{1max} is the maximum base shear associated with the inverted triangular first mode, and V_{Fmax} is the maximum shear associated with the fluctuating higher modes. V_{1max} can be taken as the static load carrying capacity under inverted triangular loading. It was established that

$$V_{Fmax} = D_m W \ddot{x}_{gmax}/g \qquad (3)$$

Where $D_m = 0.27, 0.29, 0.30$ and 0.34 for buildings with five, seven, nine, and an infinite number of stories, respectively, W is the total weight, $\ddot{x}_{gmax}$ is the peak ground acceleration, and g is the acceleration due to gravity.

Comparisons were made of calculated maximum dynamic base shears and estimates by Eqs. (2), (3). Besides the 7-story test structure, 5- and 9-story structures with the same plan were investigated, each under three different input motions. It was found that the proposed estimate provided a fair upper bound to the dynamic base shear.

The base shear associated with the first mode was carried by the wall and frames, as prescribed by the static analysis for load carrying capacity. The fluctuating shear associated with higher modes was almost exclusively carried by the wall.

ANALYTICAL INVESTIGATION OF ISOLATED SHEARWALLS

The dynamic response of isolated reinforced concrete shearwalls has been extensively investigated, as reported earlier by the author [3]. The walls represent interior elements of a building and provide, about their major axes, the entire lateral load resistance for their tributary floor areas. This is illustrated in Fig. 1. The walls were assumed to be isolated from the overall building behavior. This simplified calculation of dynamic response of walls during earthquakes.

The isolated wall was chosen not only to obtain dynamic response data for this basic element, but also to establish a reference with which to compare response of more complex wall systems. The frames in a frame-wall system or the coupling beams in a coupled-wall system are sometimes relatively flexible compared to the shearwall. In these cases, the wall can be considered as acting essentially as an isolated structural element.

Figure 1 shows an elevation of a 20-story isolated wall structure and the lumped-mass model used in its dynamic analysis. Preliminary studies indicated that it was permissible to use a model with the number of lumped masses less than the number of floors in the building. A 12-mass model was found to give results close to those obtained from a 20-mass model. Note that the concentrated masses in the lower critical region of the wall are spaced closer together than those above. This was done to obtain a more accurate picture of the deformations in this critical region.

In addition to the basic 20-story wall, 10-,30-, and 40-story structures were analyzed. Lumped mass models for the latter wall heights are shown in Fig. 2.

Dynamic analyses were carried out using the computer program DRAIN-2D[4]. The program allows for inelastic effects by allowing the formation of concentrated "point hinges" at the ends of elements where the moments equal or exceed the specified yield moments. The moment vs. end rotation characteristics of reinforced concrete beam-column elements (shearwall segments between nodes are modelled as such) can be realistically defined in terms of a basic bilinear relationship that develops into a hysteretic loop with unloading and reloading stiffnesses decreasing in loading cycles subsequent to yielding. The modified Takeda model [5] (Fig. 3), developed for reinforced concrete, was utilized in the program to represent the above characteristics.

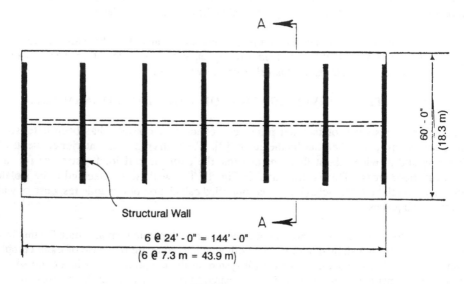

PLAN

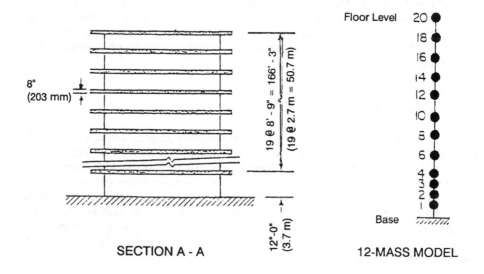

SECTION A - A

12-MASS MODEL

Figure 1. Building studied: 20-story building with isolated shearwalls

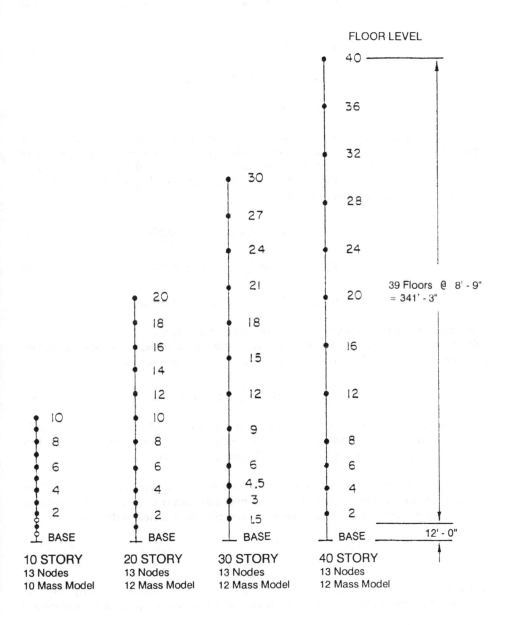

Figure 2. Lumped mass models of isolated shearwalls of various heights

Over three hundred dynamic inelastic response history analyses were carried out, with structural and ground motion parameters varied.

The maximum base shears in walls subject to earthquakes of spectral intensities* equal to 1.5, 1.0 and 0.75 times that of the 1940 El Centro, N-S record were found to be given by:

$$V_{max} = 0.127W + M_y/0.67h_n \qquad (4a)$$

$$V_{max} = 0.077W + M_y/0.67h_n \qquad (4b)$$

and
$$V_{max} = 0.060W + M_y/0.67h_n \qquad (4c)$$

respectively, where M_y is the yield moment at the base and h_n is wall height. If the spectral intensities of 1.5, 1.0 and 0.75 times that of the 1940 El Centro, N-S record were considered to be approximately equivalent to peak ground accelerations of 0.5g, 0.33g and 0.25g, respectively (the El Centro, 1940, N-S motion exhibited a peak ground acceleration of 0.33g), Eqs. 4a through c could be rewritten as:

$$V_{max} = 0.254W\ddot{x}_{gmax}/g + M_y/0.67h_n \qquad (5a)$$

$$V_{max} = 0.233W\ddot{x}_{gmax}/g + M_y/0.67h_n \qquad (5b)$$

$$V_{max} = 0.240W\ddot{x}_{gmax}/g + M_y/0.67h_n \qquad (5c)$$

respectively. Since Eqs. 5b and c were derived only for 20-story walls, while Eq. 5a applied to all wall heights, it appeared reasonable to suggest the following unified equation giving the maximum dynamic base shear in isolated walls subject to seismic excitation:

$$V_{max} = 0.25W\ddot{x}_{gmax}/g + M_y/0.67h_n \qquad (6)$$

whether the first term on the right hand side of Eq. (6) depends on wall height and/or structural period was investigated in depth. No significant dependence could be established. The term was thus left dependent solely on peak ground acceleration.

It is interesting that the proposed expression for the maximum value of the dynamic base shear is almost identical with the expression recommended earlier by Aoyama [1]. However, how much significance should be attached to this coincidence is unclear at this time.

*Spectral intensity is taken as the area under the 5%-damped relative velocity response spectrum corresponding to 10 seconds of ground motion, between periods of 0.1 second and 3.0 seconds.

It would have been of interest to record, with each maximum shear value, the corresponding state of deformation of the wall in terms of a ductility ratio or drift ratio. The absence of such deformation related information is a limitation of the above study.

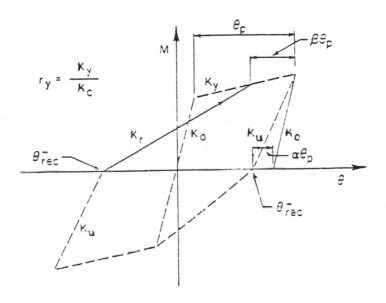

Figure 3. Modified Takeda model of moment versus end rotation characteristics of reinforced concrete members

FURTHER RESEARCH

The one-component inelastic finite beam-column element used to model shearwall segments in the above study has important limitations. The effects of inelastic shear and axial deformations are not considered. Also not considered is the fixed end rotation caused by bond slippage of the tensile reinforcement embedded in the foundation.

As part of an attempt to generalize the conclusion of the above study to include coupled-wall and frame-wall structures, a new one-component beam-column element was introduced into the program DRAIN-2D [4]. The element has the following features:

-    The existing hysteretic flexural model has been modified. A trilinear primary curve has been defined in order to input elastic cross sectional properties and to let the program determine both cracking and yielding events. Thus, the use of effective stiffness properties is avoided as input data, once an appropriate flexural primary curve is defined. The model also has been modified to consider different elastic, post-cracking and post-yielding stiffnesses for positive and negative moments. This is required for the analysis of nonsymmetric cross sections such as beams in which the slab effects are important. Also, the number of hysteretic rules have been reduced from 11 to 6.

- An additional hysteretical flexural model has been introduced following the basic characteristics of the existing model. While the existing modified Takeda model was obtained from testing of column elements, this model was obtained from testing of more than 40 walls in Japan.

- Two hysteretic shear deformation models have been introduced to concentrate inelastic shear effects at the ends of the element. The first model was developed in Canada [6] as a result of testing of column elements. The second was developed in Japan based on results of testing of wall elements [7]. From a computer programming point of view, both models were made similar, accounting for the same number of hysteretic rules. However, quite different procedures were used to define the unloading and reloading force and deformation levels where the element stiffnesses change. Also, those stiffnesses were defined in a different way. In both models realistic stiffness limitations where imposed, similar to those already included in the existing flexural model.

- An inelastic bilinear axial force-deformation model, as proposed by Kabeyasawa [8] and studied by Vulcano and Bertero [9], has been introduced. This would enable analysis of column elements when they are subjected to tension.
- A bilinear fixed-end moment vs. rotation model proposed by Filippou [10] has also been introduced. This model will take into consideration the effects of bond slippage of tensile reinforcement at structural joints for beams or at supports for columns and shear walls.

The flexural, shear and fixed-end moment vs. rotation models have been put in series within the element flexibility matrix. In addition, the computer procedure within a single time step considers the first yielding event that may occur in any of the models. Then, within the same time step, adjustments in the force, deformation and stiffness of each model are made, which will change again if a new yielding event occurs. For instance, this allows the detection and consideration of the shear reversal effect that a flexural yielding event may produce due to redistribution of internal forces.

Work still remaining includes the following:

- Establish reliable methods to determine the primary curve for each model for different structural elements (beams, columns, walls and beam-column-joints):

  • Use sectional moment-curvature analysis as basis of the flexural primary curve for all elements.

  • Use the compression field theory for determining the shear primary curve for all structural elements.

  • Establish a procedure for determining the primary curve for the fixed-end moment vs. rotation model. The curves for beams, columns and walls must be distinct from one another.

• Use the method proposed by Vulcano and Bertero [9] to define the primary curve for the inelastic axial force-deformation model.

-   Verify and calibrate the computer program including the above models by comparing analytical response with response obtained from testing columns, shearwalls, beam-column joints, and the seven-story wall-frame building studied as part of the U.S.-Japan Cooperative Program [2].

-   Compare the suitability of the one-component beam-column element incorporating the above hysteretic models with that of the three-component model proposed by Kabeyasawa [8] for the analysis of inelastic seismic response of reinforced concrete shearwalls. The model formulated by Kabeyasawa idealizes a generic wall member as three vertical line elements with infinitely rigid beams at the top and bottom floor levels: two outside truss elements represent the axial stiffness of the boundary columns, while the central element is a one-component model with vertical, horizontal and rotational springs concentrated at the base. The model is capable of describing flexural and shear deformations, while the deformation due to fixed-end rotation is not accounted for. According to Vulcano and Bertero [9], the three-vertical-line-element model can be considered to be one of the most suitable among the reinforced concrete wall models available in the literature for incorporation in a practical nonlinear analysis of multistory structural systems. It incorporates the main features of experimentally observed behavior such as migration of the neutral axis of the wall cross-section, rocking of the wall, etc., which the equivalent beam-column model fails to describe.

Conduct parametric studies of coupled-wall and frame-wall systems to generalize the conclusion of the isolated wall study concerning required shear strength of earthquake-resistant shearwalls.

CONCLUSION

Response time histories of multi-degree-of-freedom shearwall-frame systems subject to seismic input motion have indicated that the high-frequency variation of base shear due to higher mode effects is directly reflected in the wall shear. On the other hand, column shears vary in phase with the displacement time history, where the fundamental mode is predominant. This is the consequence of the fact that the wall dictates the overall displacement mode to be almost always in the inverted triangular shape. As long as this is the case, the design of frame members becomes simple. However, in order for the wall to behave in the manner indicated, it is imperative for it not to fail in shear. For the design of walls, dynamic shear which is affected by higher modes must be considered. This paper discusses the evaluation of dynamic wall shear: work completed to date as well as current and future work.

REFERENCES

1. Aoyama, H., "Earthquake Resistant Design of Reinforced Concrete Frame Building with "Flexural" Walls," Proceedings of Second U.S.-Japan Workshop on Improvement of Seismic Design and Construction Practices, ATC 15-1, Applied Technology Council, Redwood City, CA, 1987, pp. 101-129.

2. Earthquake Effects on Reinforced Concrete Structures: U.S.-Japan Research, ed. J. K. Wight, American Concrete Institute, Detroit, MI, 1985.

3. Ghosh, S. K., and Markevicius, V. P., "Design of Earthquake Resistant Shearwalls to Prevent Shear Failure," Proceedings of Fourth U.S. National Conference on Earthquake Engineering, Palm Springs, CA, V.2, Earthquake Engineering Research Institute, El Cerrito, CA, 1990, pp. 905-913.

4. Kanaan, A. E., and Powell, G. H., "A General Purpose Computer Program for Inelastic Dynamic Response of Plane Structures (DRAIN-2D), with User's Guide and Supplement," Reports No. EERC 73-6 and 73-22, University of California, Berkeley, April 1973 (Revised September 1973) and August 1975, 273 pp.

5. Takeda, T., Sozen, M. A., and Nielsen, N. N., "Reinforced Concrete Response to Simulated Earthquake," Journal of the Structural Division, ASCE, V. 96, No. ST12, December 1970, pp. 2557-2573.

6. Ozcebe, G., and Saatcioghi, M., "Hysteretic Shear Model for Reinforced Concrete Members," Journal of Structural Engineering, ASCE, V. 115, No. 1, January 1989, pp. 132-148.

7. Ono, A., Adachi, H., Nakanishi, M., and Kawakami, T., "Restoring Force Characteristics and Earthquake Response Analysis of Reinforced Concrete Shear Walls with Several Stories," Transactions of the Japan Concrete Institute, V. 6, 1984, pp. 513-520.

8. Kabeyasawa, T., "Ultimate-State Design Analysis of R/C Wall-Frame Structures," Bulletin of the Faculty of Engineers, Yokohama National University, V. 34, Yokohama, Japan, March 1985, pp. 75-100.

9. Vulcano, A., and Bertero, V. V., "Analytical Models for Predicting the Lateral Response of RC Shear Walls: Evaluation of Their Reliability," UCB/EERC-87/19, University of California, Berkeley, November 1987, 85 pp.

10. Filippou, F.C., and Issa, A., "Nonlinear Analysis of Reinforced Concrete Frames under Cyclic Load Reversals," UCB/EERC-88/12, University of California, Berkeley, September 1988, 114 pp.

MACROSCOPIC MODELING FOR NONLINEAR ANALYSIS OF RC STRUCTURAL WALLS

ALFONSO VULCANO
Associate Professor / Dipartimento di Strutture
Università della Calabria
87036 Arcavacata di Rende (Cosenza), ITALY

ABSTRACT

In this paper attention is focused on RC shear-wall models which, based on a macroscopic approach, can be efficiently incorporated in a practical nonlinear analysis of multistorey RC buildings with structural walls. An overview of the above models is given. In order to evaluate the effectiveness and reliability of selected wall models, analytical and experimental results with reference to isolated walls and to a frame-wall structure are shown.

INTRODUCTION

Reinforced Concrete (RC) shear walls are very effective in providing resistance and stiffness against lateral loads induced by earthquake and/or wind. Extensive research, both analytical and experimental [1-15], has been carried out in order to clarify, and then to simulate, the hysteretic behaviour of both isolated and coupled RC walls and of RC frame-wall structural systems. Recent research has significantly improved understanding of the inelastic behaviour of such structures, providing helpful information for the development of suitable analytical models.

Many analytical models have been proposed for predicting the nonlinear response of RC structural walls. They can be classified into two broad groups: (a) detailed models, derived using mechanics of solids, which are based on a detailed interpretation of local behaviour (microscopic approach); (b) models based on a simplified idealization, which are capable of predicting a specific overall behaviour with reasonable accuracy (macroscopic approach).

Even though the Finite Element Method offers a powerful analytical tool for simulating the nonlinear behaviour of RC structures according to the microscopic approach [1], difficulties arise because of the lack of completely reliable basic models and the complexities involved in the analysis. But, very important, the computation is generally very time-consuming and requires a large storage: thus, in practice the use of microscopic FE models is restricted to the analysis of isolated or coupled walls.

On the other hand, macroscopic wall models, because of their relative simplicity, can be efficiently adopted even for nonlinear analysis of multistorey RC structural systems. However, attention must be paid to the reliability of macroscopic models, whose validity is limited to cases such that the conditions on which the derivation of the models themselves is based would be satisfied.

In this paper the features and limitations of macroscopic wall models already proposed are considered in the light of experimentally observed behaviour. Then, suggestions are made for improving the effectiveness and/or reliability of selected wall models.

MACROSCOPIC WALL MODELS

Equivalent Beam Model

A current modeling for simulating the hysteretic behaviour of a RC structural wall considers the generic wall member replaced at its centroidal axis by a line element, which, for instance in a frame-wall structural system, is connected by rigid links to the girders. Commonly a one-component beam model is adopted. This model consists of a flexural elastic member with a nonlinear rotational spring at each end to account for the inelastic behaviour of critical regions; the fixed-end rotation at any connection interface can be taken into account by a further nonlinear rotational spring.

To simulate the propagation of the inelasticity adequately, Takayanagi and Schnobrich [2] discretized each wall member into a suitable number of short segments. However, this requires large computational effort, which could be a limitation in the nonlinear analysis of multistorey structures. Moreover, to account for the inelastic shear deformation effects in a coupled wall system, Takayanagi et al. [3] introduced additional plastic hinges at the ends of each beam element.

The main limitation of a beam model lies in the assumption that rotations occur around points of the centroidal axis of the wall. Thus, important features of the observed behaviour (i.e., fluctuation of the neutral axis of the wall cross-section, rocking, etc.) are disregarded and the consequent effects in a structural system (i.e., outriggering interaction with the frame surrounding the wall in a frame-wall structure, etc.) are not accounted for adequately.

Equivalent Truss Model

Another modeling technique represents the wall as an equivalent truss system. On the basis of experimental test results Hiraishi [4] introduced a non-prismatic truss member whose sectional area was determined according to the stress along the height of the boundary column in tension. However, the use is at the present limited to a monotonic loading, because of the difficulties in defining the structural topology and the properties of the truss elements under a cyclic loading.

Multiple-Vertical-Line-Element Models

Three-Vertical-Line-Element Model (TVLEM): This model was originally proposed by Kabeyasawa et al. [7] on the basis of the experimentally observed behaviour of a seven-storey RC frame-wall structural system. The model, shown in Fig. 1a, idealized a wall member as three vertical line elements with infinitely rigid beams at the top and bottom floor levels: two outside truss elements represented the axial stiffnesses K_1 and K_2 of the boundary columns, while the central element was a one-component model consisting of vertical, horizontal and rotational springs concentrated at the base with stiffnesses K_V, K_H and K_ϕ, respectively. The model was intended to simulate the deformation of the wall member under a uniform distribution of curvature. The hysteretic behaviour of the elements constituting the wall model was simulated by adopting empirical assumptions on the basis of the experience coming from experimental tests.

The axial-stiffness hysteresis model (ASHM) in Fig. 1b was proposed by the above authors in order to simulate the axial force-deformation relationship of the three vertical line elements. Many empirical assumptions were made. For instance, as regards the skeleton curve, when the axial force changed direction from compression to tension, the stiffness was reduced to 90% of its initial value under compression ($K_t=0.90K_c$); once tensile yielding occurred at point Y, the stiffness was reduced to 0.1% of the above initial value ($K_h=0.001K_c$). Two parameters, $\alpha=0.9$ and $\beta=0.2$, were introduced to describe the unloading stiffness degradation

K_r and the stiffness hardening point P, respectively. However, as already shown in Ref. [11], a value $\alpha < 0.687$ should be assumed in order to obtain in any case a realistic unloading path.

The origin-oriented hysteresis model (OOHM) in Fig. 1c was used for both the rotational and horizontal springs at the base of the central vertical element. The stiffness properties of the rotational spring were defined by referring to the cross-section of the central panel only, disregarding in this manner the displacement compatibility with the boundary columns. With regard to the stiffness properties of the horizontal spring, the trilinear skeleton curve was defined on the basis of empirical formulae.

Although the original TVLEM contains many empirical assumptions, its merit is that it can be considered as the early proposed macroscopic model accounting for the fluctuation of the neutral axis of the wall cross-section: this is particularly important for an adequate simulation of phenomena experimentally observed for RC frame-wall structures (e.g., outriggering effect due to the interaction of the walls with the surrounding frames).

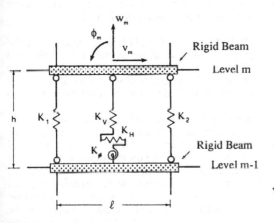

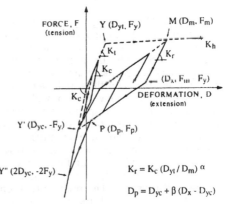

(a) Idealization of a Wall Member
   (Kabeyasawa et al. [7])

(b) Axial-Stiffness-Hysteresis Model (Ref.[7])

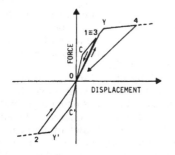

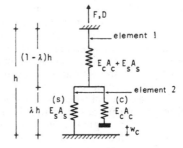

(c) Hysteresis Model for Horizontal and
   Rotational Springs in Fig. 1a (Ref. [7])

(d) Modification of the Model in Fig. 1b
   (Vulcano and Bertero [10-11])

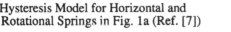

Fig. 1 - Three-Vertical-Line-Element Model and Its Modification

Modified Three-Vertical-Line-Element Model (MTVLEM): In order to limit as much as possible the empirical assumptions, in Refs. [10-11] the TVLEM was modified by replacing the ASHM with the two-axial-element-in-series model (AESM) shown in Fig. 1d: the element 1 was a one-component model to represent as a whole the axial stiffness of the column segments in which the bond was still active, while the element 2 was a two-component model to represent the axial stiffness of the remaining segments of steel (S) and cracked concrete (C) for which the bond was almost completely deteriorated. The AESM was intended to idealize the main features of the actual hysteretic behaviour of the materials and their interaction (yielding and hardening of the steel, concrete cracking, contact stresses, bond degradation, etc.). Even though refined constitutive laws could be assumed for describing the hysteretic behaviour of the materials and their interaction, in the mentioned references very simple assumptions (i.e., linearly elastic curve for the element 1, bilinear curve and linearly elastic curve in compression neglecting tensile strength, respectively for S and C components of the element 2) were made in order to carry out a first check of the effectiveness and reliability of the proposed model.

Moreover, it was observed that, in order to account for the displacement compatibility between the central panel and the boundary columns of the wall, a softening skeleton curve should be assumed for describing the response of the rotational spring of the TVLEM.

Multi-Component-in-Parallel Model (MCPM) : In Ref. [11] it was emphasized the opportunity of obtaining a more refined description of the flexural behaviour of the wall from one or both the following approaches: (a) modification of the geometry of the wall model to gradually account for the progressive yielding of the steel; (b) use of more refined laws, based on the actual behaviour of the materials and their interaction, to describe the response of the two elements in series constituting the AESM.

In Ref. [12] a new wall model was proposed by following both the above approaches (Fig. 2). The flexural response of a wall member was simulated by a multi-uniaxial-element-in-parallel model with infinitely rigid beams at the top and bottom floor levels: the two external elements represented the axial stiffnesses K_1 and K_2 of the boundary columns, while two or more interior elements, with axial stiffnesses $K_3,....., K_n$, represented on the whole the axial and flexural stiffnesses of the central panel. A horizontal spring, with stiffness K_H and hysteretic behaviour described by the OOHM already mentioned, simulated the shear response of the wall member. The relative rotation $\Delta\phi_m$ was intended around the point placed on the central axis of the wall member at height ch. A suitable value of the parameter c could be selected on the basis of the expected curvature distribution along the inter-storey height h: for instance, $0<c<1$, if the curvature presented tha same sign along h.

A modified version of the AESM above shown was proposed to describe the response of a uniaxial element (Fig. 2b). Analogously to the original AESM, the two elements in series were still representative of the axial stiffness of the column segments in which the bond remained active (element 1) and those segments for which the bond stresses were negligible (element 2). Unlike the AESM, also the element 1 consisted of two parallel components to account for the mechanical behaviour of the uncracked concrete (C) and the steel (S); a suitable law for the dimensionless parameter λ defining the length of the two elements provided with an accurate description of the measured tension-stiffening effect. Refined constitutive laws were adopted to idealize the hysteretic behaviour of the materials and the tension-stiffening effect.

Exactly, the stress-strain relationship proposed by Bolong et al. [16] was assumed for cracked concrete, because accounting for the contact stresses due to the progressive closure of cracks (Fig. 3a). The stress-strain relationship in Fig. 3b and a set of rules under a generalized load history, both proposed in Ref. [12], were adopted for uncracked concrete. The stress-strain relationship originally proposed by Giuffrè and Pinto [17] and later implemented by Menegotto and Pinto [18] was adopted to describe the hysteretic response of the reinforcing steel (Fig. 3c); in order to avoid the storage of all parameters required for a generalized load history to retrace all previous reloading curves which were left incomplete, a set of simple rules was used.

Under monotonic tensile loading the tension-stiffening effect was taken into account by calculating the value of the dimensionless parameter λ such that the tensile stiffness of the uniaxial model in Fig. 2b would be equal to the actual tensile stiffness of the uniaxial RC member which was intended to be idealized:

$$\left\{ \frac{(1-\lambda)\,h}{E_{ct}\,A_c + E_s\,A_s} + \frac{\lambda\,h}{E_s\,A_s} \right\}^{-1} = \frac{E_s\,A_s}{h}\,\frac{\varepsilon_s}{\varepsilon_m}$$

where $E_{ct}A_c$ and E_sA_s are the axial stiffnesses in tension of the concrete and of the steel, respectively, while the ratio $\varepsilon_s/\varepsilon_m$ of the steel strain in a cracked section to the current average strain for the overall member was evaluated by the empirical law proposed by Rizkalla and Hwang [19]. Under cyclic loading it was assumed that, along an unloading path from a tensile stress state, the value of λ was kept constantly equal to the value corresponding to the maximum tensile strain previously attained; if this maximum strain was exceeded during a tensile reloading, the value of λ was updated as for the case of monotonic tensile loading.

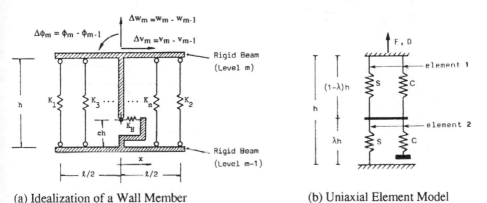

(a) Idealization of a Wall Member   (b) Uniaxial Element Model

Fig. 2 - Multi-Component-in-Parallel Model (Vulcano et al. [12])

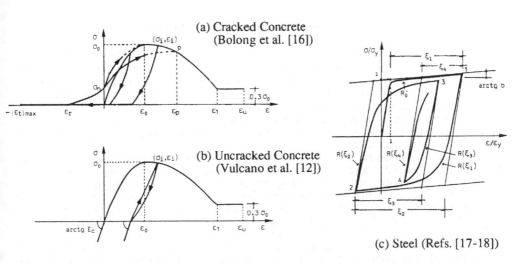

Fig. 3 - Material Constitutive Laws Adopted for Model Components in Fig. 2b

Modified Multi-Component-in-Parallel Model (MMCPM): As it will be shown in the next section, the accuracy in predicting the flexural response of the wall by the MCPM is excellent when the constitutive laws in Fig. 3 are adopted for the modified-AESM components in Fig. 2b, even by assuming few uniaxial elements. However, the above constitutive laws are very sophisticated; to improve the effectiveness of MCPM without renouncing a reasonable accuracy, the use of simplified (more schematic) constitutive laws could be suitable as well. This is confirmed in recent studies conducted by Fajfar and Fischinger [14], who introduced simplified hysteretic rules to describe the response of both the vertical and horizontal springs (Fig. 4). A further variant of the MCPM was applied in a study of Fischinger at al. [15], who, in order to reduce the uncertainty in the assumption of a suitable value for the parameter c, used a stack of many elements which were placed one upon the other.

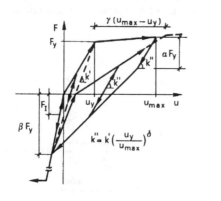

(a) Vertical Spring Rules (b) Horizontal Spring Rules

Fig. 4 - Modified Hysteretic Behaviour Rules for Springs of the Wall Model in Fig. 2a
(Fajfar and Fischinger [14], Fischinger et al.[15])

RESPONSE OF MACROSCOPIC MODELS

In order to check the effectiveness and reliability of macroscopic wall models, in the course of previous studies [10-13] an extensive numerical investigation was carried out with reference to isolated walls tested under monotonic and cyclic loadings by Vallenas et al. [5] at the University of California at Berkeley (UCB) (Fig. 5) and to a 7-storey RC frame wall structure, which was tested, within the framework of a joint U.S.-Japan research project [6], at Tsukuba (full-scale pseudo-dinamic testing) by Kabeyasawa et al. [7] and at the UCB (1/5th-scale dynamic testing, on earthquake simulator) by Bertero et al. [8] (Fig. 6). Herein some meaningful results are shown and critically discussed.

In Figs. 7a and 7b analytical results obtained by adopting, respectively, the TVLEM and the MTVLEM, are compared with the experimental results. Exactly, the analytical curves in Fig. 7a have been obtained by assuming for the degradation parameter α the value 0.9 suggested in Ref. [7] and the limit value 0.687. Both the analytical curves, but particularly that corresponding to α=0.9, exhibit mechanical degradation. On the contrary, the improvement obtained by the MTVLEM, even by assuming very simplified laws for the AESM components in Fig. 1d, is evident, also in comparison with the response obtained by a F.E. analysis.

A considerable improvement in the prediction of the wall flexural behaviour can be attained by the MCPM. This is clearly shown in Figs. 8a and 8b, in which analytical and experimental curves are compared for specimens 5 and 6 in Fig. 5b, subjected to monotonic and cyclic loadings, respectively. All the analytical curves represented by a full line in Fig. 8a have been obtained by assuming 4 vertical uniaxial elements (n=4) and different values of the parameter c. Apart from a slight discrepancy, the correlation of the experimental curve and the analytical curve corresponding to c=0.4 can be considered good; it becomes excellent when adopting the values c=0.4 and n=8 (see dashed line). The correlation is very good also with reference to the curves in Fig. 8b, where the analytical curve represents the displacement-controlled flexural response. Analogous results, which are omitted for sake of brevity, have been obtained for specimens 3 and 4 in Fig. 5a. However, it should be noted that under high shear stresses, likewise TVLEM, difficulties are met for accurately describing by MCPM experimentally observed flexural and shear displacement components. This is clearly shown in Fig. 8c, with reference to specimen 5, by assuming different values of the parameter c and a relatively detailed discretization (n=8) .

Finally, in order to show the suitability of the MCPM for incorporation in the nonlinear analysis of multistorey structures, the test structure in Fig. 6 has been modeled as shown in Fig. 9, that is by idealizing all the structural members as line beam elements (LGM model) or, alternatively, the wall only by the MCPM (WGM model); a suitably reduced value of the initial flexural stiffness of the beam elements, idealized by a two-component model, has been assumed. In Fig. 10 experimental and analytical curves are compared with reference to the above structure. Exactly, in Figs. 10a and 10b shears (V_T, V_w) and overturning moments (M_T, M_w), respectively, evaluated at the base of the overall structure or of the wall are shown against the roof horizontal displacement v_7. The experimental curves were obtained by Bertero et al. [8] as envelope of the results of dynamic testings, whereas the analytical curves have been obtained by a static nonlinear analysis for a uniform loading distribution.

As it can be observed, both the analytical models, but particularly LGM, underestimated the ultimate strength of the structure, even though a uniform loading distribution has been assumed for both of them: this, as noted in Ref. [13], should imply some overestimation of the ultimate strength, if the actual loading distribution in the dynamic testing and some mechanical degradation due to cyclic loading are taken into account. On the other hand, the underestimation of the ultimate strength by both the analytical models can be mostly ascribed to the fact that the strain rate to obtain the constitutive curves of the materials, on which the mechanical characteristics of the analytical models have been settled, was considerably lower than that recorded during the dynamic testing. However, it should be noted that the LGM model, in which the rotation of the central wall is described as occurring around the axis of the wall itself, gives rise to a wrong description of the rocking and consequent spatial-interaction effects. Thus, an appreciable increase of the axial compressive force in the central wall has been observed when using WGM model, whereas the axial force for the wall remained practically equal to the gravity axial force when using the LGM model. A further consequence has been that very different values of the ductility demand have been obtained for both longitudinal and transverse girders by adopting the two analytical models.

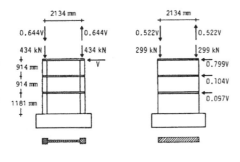

(a) Specimens 3,4   (b) Specimens 5,6

Fig. 5 - Test Walls (Vallenas et al. [5])

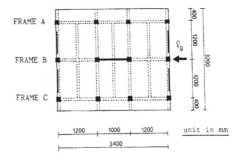

Fig. 6 - Test Structure (Bertero et al. [8])

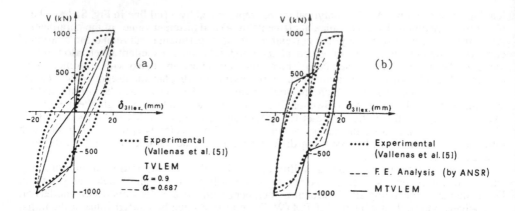

Fig. 7 - Analytical and Experimental Curves for Specimen 4

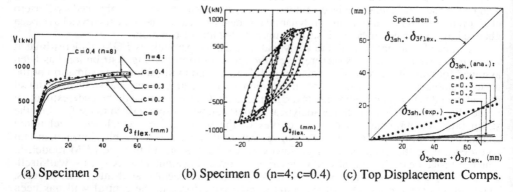

(a) Specimen 5   (b) Specimen 6 (n=4; c=0.4)   (c) Top Displacement Comps.

Fig. 8 - Analytical and Experimental Curves for Test Walls in Fig. 5b

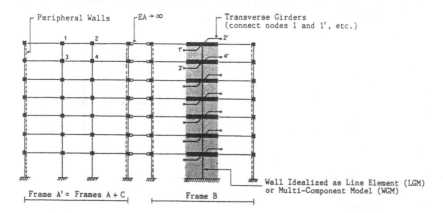

Fig. 9 - Pseudo-Three-Dimensional Model of the Test Structure in Fig. 6

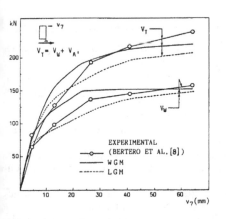

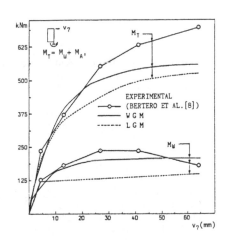

(a) Base Shears for Overall Structure and Wall Vs. Roof Displacement

(b) Base Overturning Moments for Overall Structure and Wall Vs. Roof Displacement

Fig. 10 - Analytical and Experimental Curves for Test Structure in Fig. 6

CONCLUSIONS

Features and limitations of models idealizing a RC structural wall have been discussed in the previous sections. As mentioned, wall models based on a macroscopic approach are more effective than microscopic FE models for the purpose of incorporation in the nonlinear analysis of multistorey structures. Equivalent beam and truss models present many limitations. Indeed, a beam model, although suitable for its simplicity, is not capable of simulating the fluctuation of the neutral axis of the wall cross-section, giving rise to a wrong description of the interaction with the other structural members. On the other hand, the implementation of a truss model meets with difficulties in defining the properties of the truss elements, particularly under cyclic loading.

Multiple-vertical-line-element models prove to be the most suitable for the purpose mentioned above, particularly because they are capable of accounting for fluctuation of the cross-section neutral axis. As shown, a multi-component-in-parallel model accurately predicts the flexural response, even assuming the minimum of uniaxial elements (n=4) with the advantage of a limited computational effort. The accuracy in predicting the flexural response depends, rather than assuming a greater number of uniaxial elements, on the choice of a suitable value of the parameter c according to the expected distribution of curvature along the wall member. By assuming c=0.4 for the test walls examined in this paper and refined constitutive laws for the uniaxial elements, the correlation of analytical and experimental results has been excellent. In order to improve the effectiveness of the model, the use of simplified, yet reasonably accurate, constitutive laws can be adequate as well; moreover, the uncertainty in defining the parameter c can be overcome by placing more elements one upon the others.

However, under high shear stresses, the OOHM gives only approximate description of the shear hysteretic response; thus, the prediction of the shear and flexural displacement components becomes difficult. Therefore, improvement of the wall model is needed by revising the OOHM and/or by introducing some relation between flexural and shear responses, which have been independently described. Further improvements can be pursued by making the wall model capable of simulating other observed phenomena at the present ignored (e.g., fixed-end-rotation at the base of the wall, etc.) and by a better calibration of the parameters affecting the response of the models (e.g., c) on the basis of integrated analytical and experimental research.

REFERENCES

1. American Society of Civil Engineering, "Finite Element Analysis of Reinforced Concrete", State-of-the-Art Report, ASCE, New York, 1982.
2. Takayanagi T. and Schnobrich W.C., "Nonlinear analysis of coupled wall systems", Earthquake Engineering and Structural Dynamics", Vol. 7, 1979.
3. Takayanagi T., Derecho A.T. and Corley W.G., "Analysis of inelastic shear deformation effects in reinforced concrete structural wall systems", Nonlinear Design of Concrete Structures, CSCE-ASCE-ACI-CEB Int. Symp., University of Waterloo, Canada, 1979.
4. Hiraishi H., "Evaluation of shear and flexural deformations of flexural type shear walls", Procs. 4th Joint Tech. Coord. Committee, U.S.-Japan Coop. Earth. Research Program, Building Research Institute, Tsukuba, Japan, 1983.
5. Vallenas J.M., Bertero V.V. and Popov E.P., "Hysteretic behaviour of reinforced concrete structural walls", Report No. UCB/EERC-79/20, Earthquake Engineering Research Center, University of California, Berkeley, 1979.
6. U.S. Members of JTCC Group on RC Building Structures.,"U.S.-Japan research: seismic design implications", J. of Struct. Eng., ASCE, Vol.114, No.9, Sept.1988.
7. Kabeyasawa T., Shioara H. and Otani S., "U.S.-Japan cooperative research on R/C full-scale building test - Part 5: discussion on dynamic response system", Procs. 8th W.C.E.E., Vol. 6, S. Francisco, 1984.
8. Bertero V.V., Aktan A.E., Charney F.A. and Sause R., "U.S.-Japan cooperative research program: earthquake simulation tests and associated studies of a 1/5th-scale model of a 7-story reinforced concrete test structure", Report No. UCB/EERC-84/05, University of California, Berkeley, 1984.
9. Fajfar P. and Fischinger M., "Nonlinear seismic analysis of RC buildings: implications of a case study", European Earthquake Engineering, No. 1, 1987.
10. Vulcano A. and Bertero V.V., "Nonlinear analysis of R/C structural walls", Procs. 8th E.C.E.E., Vol. 3, Lisbon, 1986.
11. Vulcano A., and Bertero V.V., "Analytical models for predicting the lateral response of RC shear walls: evaluation of their reliability", Report No. UCB/EERC-87/19, University of California, Berkeley, 1987.
12. Vulcano A., Bertero V.V. and Colotti V., "Analytical modeling of R/C structural walls", Procs. 9th W.C.E.E.,Vol. VI, Tokyo-Kyoto, Japan, 1988. See also: Colotti V. and Vulcano A., "Behaviour of RC structural walls subjected to large cyclic loads" (in Italian), Procs. Giornate A.I.C.A.P. (Italian Association of Reinforced and Prestressed Concrete Structures), Vol. 1, Stresa, Italy, 1987.
13. Vulcano A. and Colotti V., "Analytical modeling of RC frame-wall structural systems", Procs. 9th E.C.E.E., Vol. 10-B, Moscow, 1990.
14. Fajfar P. and Fischinger M., "Mathematical modeling of reinforced concrete structural walls for nonlinear seismic analysis", EURODYN'90, Euro. Conf. on Struct. Dyn., Bochum, Germany, 1990.
15. Fischinger M., Vidic T., Selih J., Fajfar P., Zhang H.Y. and Damianic', "Validation of a macroscopic model for cyclic response prediction of R.C. walls", Procs. 2nd Int. Conf. on Computer Aided Analysis and Design of Concrete Structures, Zell am See, 1990.
16. Bolong Z., Mingshun W. and Kunlian Z., "A study of hysteretic curve of reinforced concrete members under cyclic loading", Procs. 7th W.C.E.E., Vol. 6, Instanbul, Turkey, 1980.
17. Giuffrè A. and Pinto P.E., "The behaviour of reinforced concrete under strong-intensity cyclic loading" (in Italian), Giornale del Genio Civile, N.5, 1970.
18. Menegotto M. and Pinto P.E., "Method of analysis for ciclycally loaded reinforced concrete plane frames including changes in geometry and nonelastic behavior of elements under combined normal force and bending", Procs. IABSE Symp. on Resistance and Ultimate Deformability of Structures Acted on by Well-Defined Repeated Loads, Lisbon, 1973.
19. Rizkalla S.H. and Hwang L.S., "Crack prediction for members in uniaxial tension", ACI Journal, Nov.-Dec., 1984.

NONLINEAR SEISMIC ANALYSIS OF STRUCTURAL WALLS USING THE MULTIPLE-VERTICAL-LINE-ELEMENT MODEL

MATEJ FISCHINGER, TOMAŽ VIDIC, PETER FAJFAR
University of Ljubljana, Department of Civil Engineering,
Institute of Structural and Earthquake Engineering,
Jamova 2, 61000 Ljubljana, Slovenia

ABSTRACT

The multiple-vertical-line-element model (MVLEM) has been applied to the inelastic static and dynamic response analysis of structural walls. Three different structural systems (a cantilever wall, a frame-wall building, and a coupled wall) were selected as illustrative examples. A comparison of analytical and experimental results indicates that the MVLEM can successfully predict the inelastic behaviour of all three different types of structural walls. The main advantage of this model is its ability to simulate the shift of the neutral axis (due to the lifting of the tension edge of the wall after yielding) and to take into account the influence of a fluctuating axial force on the strength and stiffness of the wall. Both of these features are particularly important in the case of coupled wall systems. There are, however, several problems which need further investigation. They include (i) the modeling of inelastic shear behaviour, (ii) refinement of the models for vertical springs, (iii) the calibration of the model parameters and (iv) the formulation of a new model for coupling beams.

INTRODUCTION

Properly designed and correctly constructed reinforced concrete (RC) structural walls may, in addition to their high strength, exhibit very high ductility. Due to their stiffness, they reduce the seismic damages of non-structural systems. Consequently, they are very effective in providing safe and sound structural systems in earthquake regions. Unfortunately, the present capability for the realistic and practical mathematical modeling of the nonlinear static and dynamic seismic response of RC structural walls is limited. Among many proposed models, which have been discussed for example in [1], the multiple-vertical-line-element model (MVLEM) has gained a lot of interest lately, and the model is believed to be reasonably reliable in the seismic analysis of RC buildings with structural walls (e.g. [2,3]). In the paper, the basic concepts of the MVLEM are discussed and its effectiveness has been analysed by illustrative examples for three different types of structural walls.

THE MULTIPLE-VERTICAL-LINE-ELEMENT MODEL (MVLEM)

The physical model

Following the full-scale test carried out on a seven story RC frame-wall building in Tsukuba (U.S. - Japan program [4]), Kabeyasawa et al [2] proposed a new macroscopic three-vertical-line-element model (Fig. 1). In the model three vertical elements are connected by rigid beams at the top and bottom floor levels. Two outside truss (uniaxial) elements represent the axial stiffness of the boundary columns. The central vertical element, representing the panel of the wall, is a one-component model in which the vertical, horizontal and rotational springs are concentrated at the base. Since it is difficult to assign justifiable values to the rotational spring, a modified model (Fig. 2) was proposed by Vulcano et al [3]. The rotational spring was replaced by several (N) parallel vertical truss elements, which represent the axial and flexural stiffness of the central panel. The horizontal spring, which models the shear behaviour of the wall member, has remained in the model. This model is called the multiple-vertical-line-element model (MVLEM). It was incorporated into the DRAIN-2D program by the authors and used in the present study.

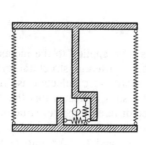

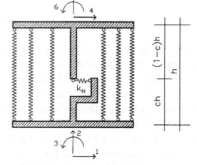

Figure 1. Three-vertical-line-element model (TVLEM) [2]   Figure 2. Multiple-vertical-line-element model (MVLEM) [3]

The entire wall is modeled as a stack of n MVLEM wall elements which are placed one upon each other. The flexural and shear deformations are separated in each MVLEM (Fig. 3). All shear behaviour is concentrated in the horizontal spring with stiffness k_H, which is placed at the height ch ($0 \leq c \leq 1$). The horizontal shear displacement at the top of the stack does not depend on c. Flexural deformations, however, do depend on c, as well as on n. The parameter c defines the relative rotation between the top and bottom levels of the MVLEM (Fig. 3). If moment (curvature) distribution along the height of the element is constant, $c = 0.5$ yields "exact" rotations and displacements for elastic and inelastic behaviour. For a triangular distribution of bending moments, $c = 0.5$ still yields exact results for rotations in elastic range, but it underestimates displacements. This problem can be solved by the stacking of elements, which leads to a small moment gradient. In the inelastic range, however, the problem is more critical, since even small moment gradients can cause highly nonlinear distributions of curvature. Consequently, lower values of c should be used to take into account the non-linear distribution of curvature along the height of the wall. For the analysed multistory walls, good results were obtained if: (i) 3 to 4 MVLEM's were used in the potential hinge area (the first story) and $c = 0.3$ was chosen.

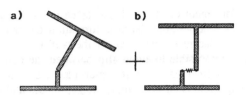

Figure 3. Flexural (a) and shear (b) deformations of the MVLEM

Hysteretic models

Based on the results of the test in Tsukuba, Kabeyasawa [2] proposed a special axial force - deformation formulation for vertical springs. Similar, though simplified, hysteretic rules, which had been originally developed by the authors for the contact elements in the joints of large panel buildings, were used in the present examples (Fig. 4). Modeling of inelastic shear behaviour has not yet been appropriately solved. Simplified rules were included into the MVLEM by the authors (Fig. 5). In the present study, inelastic shear behaviour was considered in the case of the cantilever wall only.

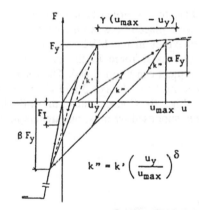

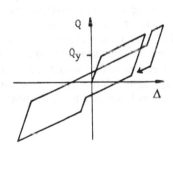

Figure 4. Vertical spring behaviour   Figure 5. Horizontal spring behaviour

THE MODELING OF COUPLING BEAMS

The response of a coupled wall depends strongly on the behaviour of the coupling beams, which is specific and different from that of beams in frames. The typical behaviour of a conventionally reinforced coupling beam is shown in Fig. 6. During the first half cycle a crack opens on one side (e.g. at the top) (Fig. 6a). During the second half cycle it may happen that the crack at the bottom opens before the one on top closes. In such a case a gap spreads over the entire height of the coupling beam (Fig. 6b), which is associated with moment capacity reduction and shear-slip. As well as this, the lugs of the deformed reinforcement cause localized crushing of the concrete in the first half cycle. This would cause the development of the voids behind the lugs. When the direction of loading in the beam is reversed, the reinforcement must slip the distance of the voids before the lugs can bear on their opposite faces. Furthermore, large deformation demands are imposed on the beams after yielding of the adjacent tension edge of the wall pier which tends to uplift. In such a case it is practically impossible to avoid slip between the conventionally reinforced beam (without diagonal reinforcement) and the pier (see also Fig. 14). Finally, if the deformation in the reloading cycle is large enough, the gap closes and the stiffness increases (Fig. 6c).

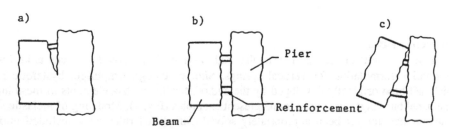

Figure 6. Behaviour of a coupling beam: (a) first crack, (b) crack opens over the entire height, (c) gap closure

A simple model has been proposed to simulate the observed behaviour (Fig. 7a). Springs following shear-slip hysteretic rules (Fig.7c) were placed between the beam and the piers (note that similar were used to model shear behaviour in the MVLEM). With these springs, the level of the shear force in the beam (the axial force in the piers) can be controlled and the slip associated with the uplift of the piers can be modeled. The rotational degrees of freedom in the contacts between the springs and beam (nodes 3 and 4) are fixed, which results in the deformation mode illustrated in Fig. 7b. Consequently, the inflexion point still occurs in the middle of the beam, as has been supposed by most researchers in the past.

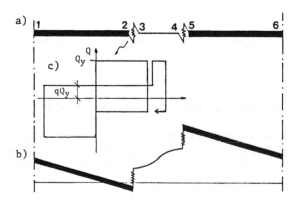

Figure 7. Proposed model for coupling beams: (a) undeformed configuration, (b) deformed configuration, (c) hysteretic rules for the shear springs

EXAMPLE No. 1: A CANTILEVER WALL

The analytical results [1] were compared with the results of a test on a simple cantilever wall with a rectangular cross-section (Fig. 8). The test was conducted at Tsinghua university in Beijing. The height of the wall was 2.4 m, the cube strength of the concrete was 36.0 MPa and the yield strength of reinforcement was either 381 MPa (ϕ 12 mm bars) or 288 MPa (ϕ 8 mm bars). The wall was subjected to cyclic loading. A MVLEM was used in the analysis. The analytical and test results agreed well for the following set of parameters: n = 8, N = 6, c = 0.3, α = 1.0, γ = 1.05 and δ = 0.5. The response was very much influenced by the parameter β, which determined the "fatness" of the hysteresis of the vertical springs. It had to be defined by a trial and error procedure. The relation $\beta = 1.5 + F_I / F_y$ was adopted, where F_I and F_y were the initial compression force and the yield force in the spring, respectively. For comparison two other mathematical models were used: (i) a simple equivalent beam model and (ii) a finite element model (FEM) [5]. The following was concluded:

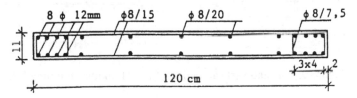

Figure 8. Cross-section of the cantilever wall

1. As far as the global response relation top displacement - base shear is concerned, the simple beam element simulated the response as successfully as the MVLEM. This is not surprising, since the moment - rotation relationship for flexural behaviour of the beam element can be relatively reliably determined (in fact more approximations were used in the case of the MVLEM).
2. A larger discrepancy was observed in the case of the FEM. One reason for this might be the perfect bond between the steel and concrete, which was enforced in the model. This proves once again that FE models often depend on parameters which are difficult to define or control. It should be noted, however, that in the MVLEM the parameter β was adjusted while the parameters in the FEM were not.

EXAMPLE No. 2: A 7-STORY FRAME-WALL RC BUILDING

The well-known 7-story RC dual building tested at full scale in Tsukuba (Japan) offers an ideal example to test the efficiency of the chosen mathematical model. The floor-plan and cross-section of the building are shown in Fig. 9. A detailed description of the building has been published in many reports (e.g. [4]) and so will not be repeated here.

The building was analysed by the authors of this paper by using a simple equivalent beam element for the structural wall. Fair correlation of the measured and computed overall response of the structure, expressed, for example, in terms of the top displacement time-history, was observed (Fig. 10) [6]. The correlation of the detailed response, however, was not so favourable. A large discrepancy was observed in the lower part of the displacement envelope (Fig. 11), where the measured displacements were much larger than the computed ones. According to the analysis, the columns did not yield at the base, although the results of the test indicated plastic hinges at the column base. In addition to this, some parameters of the beam element model (e.g. the hardening slope) had to be based on the test results.

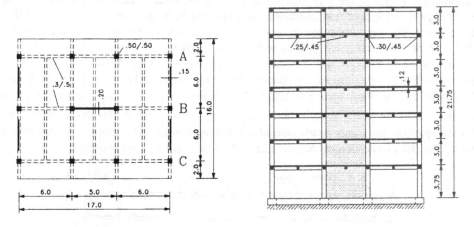

Figure 9. Plan and vertical section of the "Tsukuba" building [4].

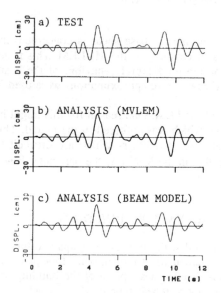

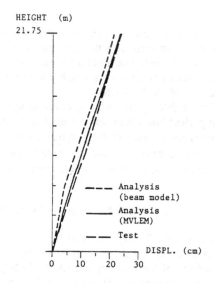

Figure 10. Top displacement time history for the "Tsukuba" building: (a) test, (b) MVLEM, (c) beam model

Figure 11. Displacement envelope for the "Tsukuba" building: Comparison of test and analyses

The inelastic response of the same building to the PSD-3 loading was recalculated by using the MVLEM for the structural wall [7]. Three elements were used in the first floor and one element in each of the other floors. Six vertical springs were employed in all the elements. The location of the horizontal spring was defined by the parameter c = 0.3. Elastic behaviour of all shear springs was assumed. The following parameters were used to control the response of the wall elements: $\alpha = \beta = \gamma = 1.0$ and $\delta = 0.5$. The strain hardening ratio of the beam vertical springs in tension was 0.01. The modeling of columns and beams was, with a small exception of beams subjected to a negative bending moment (tension at the top), the same as that for previous models in the analysis. Additional springs were used to simulate the three-dimensional behaviour of the building (the influence of the transverse beams). The accelerogram which was applied in Test PSD-3 in Tsukuba was based on the E-W component of the Taft record (1952). The maximum ground acceleration was 320 cm/s^2.

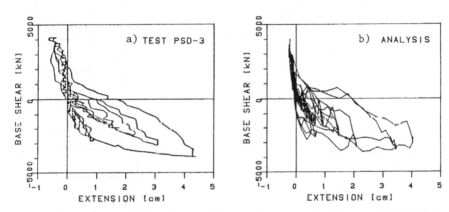

Figure 12. Extension of the right boundary column in the first story of the "Tsukuba" building: (a) test, (b) MVLEM

Some results of the analysis are shown in Figs. 10-12. The correlation between the measured and computed values is favourable not only in the case of the top displacement time-history (Fig. 10), but also in the case of the axial deformation of the boundary column of the wall (Fig. 12) and of the building's displacements envelope (Fig. 11). The results have proved that the MVLEM was capable of simulating the detailed response of the wall, which controlled the overall response of the tested building. Note also that some differences arise from the fact that MDOF model was used in the analysis, while SDOF response had been enforced in the test. The same failure mechanism as that observed in Tsukuba was predicted.

EXAMPLE No. 3: A COUPLED WALL

A 6-story coupled structural wall (Fig. 13), which was tested by Lybas and Sozen [8], was chosen as the third example. The choice was partly based on the excellent documentation of the results as well as of the input data. A total of six small-scale structures were tested. Five test structures were subjected to the scaled El Centro NS motion and one structure (speciment S1) was subjected to cyclic lateral loads. For example, the crack pattern after the cyclic test is shown in Fig. 14.

Only the preliminary results for specimen S1 (referred to as the "coupled wall" in the following text) are given below. Lybas and Sozen used a simple beam element, which disregarded axial force - flexural interaction, to model the piers of the coupled wall. An

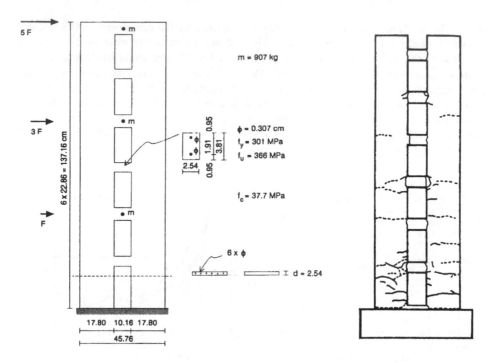

Figure 13. The coupled wall

Figure 14. Crack pattern after the cyclic test [8]

extensive parametric study was performed to define a suitable model for the coupling beams. It was necessary to modify the original Takeda rules to account for shear-slip, the slip of the longitudinal reinforcement in the beams, and gap closure in the reloading cycle. A line element which included axial force - flexural interaction was used by Keshavarzian and Schnobrich [9] to model the wall piers of the same coupled wall. They reported that the interaction did not affect the overall response (stiffness) of the wall. However, yielding of the tension pier did change the distribution of the bending moment between the two piers, as well as the distribution of shear. In the present study, both, modified beam behaviour and axial force - flexural interaction have been taken into account. In addition, the uplift of the tension edges of the piers has been considered.

Four different models (A - D) were used in the analysis and the results were compared with the experimentally observed top level load - top level lateral deflection relation of the wall when subjected to cyclic lateral loads (Fig. 15e). In all the models four MVLEM's were used in the first story of both piers and one MVLEM was used in each of all the other stories. The parameters $N = 6$, $c = 0.3$, $\alpha = 1.0$, $\gamma = 1.0$ and $\delta = 0.5$ were used throughout the analysis. One percent of hardening after yielding was assumed for the vertical springs in the MVLEM. Details of the individual models are discussed below.

Model A
1. The coupling beams were modeled as simple beams following the original Takeda rules.
2. The moment - rotation envelope for the beams was based on the assumed antisymmetric distribution of moment and curvature. The yield moment (M_y = 5.6 kNm) and ultimate moment (M_u = 7.2 kNm) were calculated. The calculated ultimate moment was lower than that reported in [8] (8.81 kNm).
3. The calculated hardening slope for the beams was 2.7 %.
4. β = 1.0 was chosen in the MVELM.
5. The compressive axial stress in the vertical springs of the MVLEM was limited to 32 MPa = 0.85 f_c.

The correlation between the test and calculated results is poor (Fig. 15a). Large rotational ductility demand (μ_Θ = 90) was imposed on the coupling beams after the uplift of the adjacent tension edge of piers, which amounted to 0.5 cm (6 cm in the prototype structure). In reality, beams cannot sustain such a large ductility demand. In the applied model, however, the ductility was not limited and, due to the hardening in the post yield range, the bending moment in the beams increased up to 19.0 kNm. Consequently, the shear forces in the beams, the axial forces in the piers, the flexural capacity of the wall and the horizontal resistance of the wall increased unrealistically. Finally, the vertical springs of the MVLEM at the base yielded in compression. Detail "x" in the cyclic response (Fig. 15a) can be explained by an increase in the stiffness of the vertical springs which occurred after the force had fallen below the yield level again. The shear forces in the beams could be limited if a smaller hardening ratio in the beams was used. This "solution", however, has no physical background.

Model B
While all the other parameters were kept constant, the proposed modified beam (Fig. 7) was used. The yield force of the shear spring was determined from the ultimate moment calculated for the coupling beam (Q_y = 2 M_u / L_b, where L_b is the clear length of the beam). No degradation of the force in the shear springs was assumed (q = 1.0, Fig. 7c). The following can be observed (Fig. 15b):

1. In comparison with Model A, much better agreement was achieved.
2. The calculated horizontal force is lower than was observed during the test. However, if the (higher) values for the ultimate moment in beams, which had been reported in [8], were used, the correlation would be better.
3. A sudden increase in stiffness on the reloading branch (detail "xx"), which was not observed during the experiment, can be noted. This problem was further investigated by the other two models.

Model C
Model C was the same as Model B, except for the degradation of the force in the shear springs (q = 0.3). The general shape of the hysteresis loops was improved (Fig. 15c). Nevertheless, the mentioned stiffness increase persisted in them.

Model D
While all the other parameters of Model C were kept constant, the parameter β of the vertical springs in the MVLEM was determined according to the relation which had been proposed for the cantilever wall in Example 1 (β =1.5 + F_I / F_y = 2.5). The observed

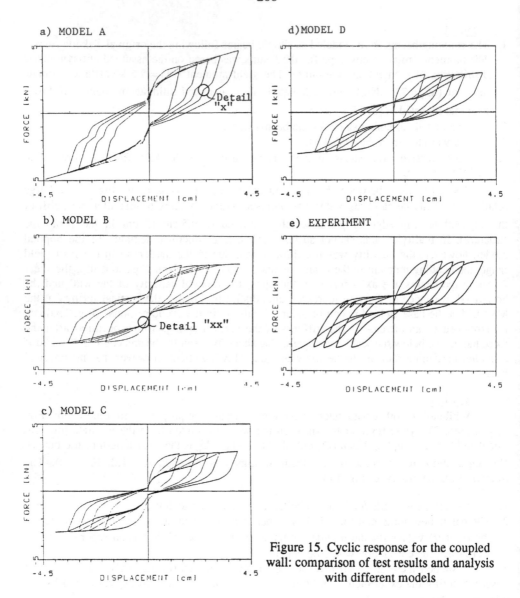

Figure 15. Cyclic response for the coupled wall: comparison of test results and analysis with different models

increase of stiffness vanished from the cyclic response hysteresis (Fig. 15d) and reasonable agreement with the test results was achieved. Noticable strength degradation, which was not successfully simulated, might be controlled by a larger γ value in the model for the vertical springs. However, this has not yet been verified.

CONCLUSIONS

1. The MVLEM quite successfully balances the simplicity of a macroscopic model and the refinements of a microscopic model. Its physical concept is clear, and the computational effort needed is reasonable. It enables modeling of some important features (e.g. shift of the neutral axis, the effect of a fluctuating axial force, inelastic shear behaviour) which have been frequently ignored in previous seismic analyses of structural walls.
2. The MVLEM was able to predict the inelastic static, cyclic and (in one case) dynamic behaviour of three different types of structural walls (a cantilever wall, a coupled wall and a frame-wall structural system).
3. While there was no particular advantage of the MVLEM over a simple beam model in simulating the global response (top displacement - base shear) of the cantilever wall, the advantages of the MVLEM were clearly seen in the case of the other two structural systems, where structural walls were connected with other elements which restrained their local deformations.
4. Successful simulation of the frame-wall interaction had been expected. After all, the basic concept of the model was proposed in accordance with the test results obtained for the analysed frame-wall building.
5. It is a new observation, however, that the MVLEM is particularly suitable for modeling coupled wall response. Moreover, the authors are convinced that a realistic estimate of the demand in coupling beams is not possible if the shift of the neutral axis and the influence of the fluctuating axial forces in the walls are not properly taken into account.
6. Vice versa, the behaviour of coupling beams controlled the response of the analysed coupled wall. Simple beam model, following the original Takeda hysteretic rules, was found inadequate to predict this behaviour at the large displacements imposed on the wall, associated with large vertical elongations of the tension edges of the coupled piers. A new model for coupling beams is therefore needed. In the presented analysis shear slip elements were added to the interfaces of the coupling beams and wall piers.
7. The behaviour of all the analysed walls was predominantly flexural and the level of the axial forces was relatively low. To account for the highly inelastic shear behaviour better models are needed. If the level of compressive axial forces was higher, nonlinear behaviour (including the confining effect) of the vertical springs in compression should be taken into account.
8. Some of the model parameters (in particular c and β) have an important influence on the response. Recommendations for their values are given in the paper. They need, however, further calibration.

ACKNOWLEDGEMENTS

The results presented in this paper are based on work supported by the Ministry of Research of the Republic of Slovenia and by the U.S. - Yugoslav Joint Fund for Scientific and Technological Cooperation.

REFERENCES

1. Fajfar, P. and Fischinger, M., Mathematical modeling of reinforced concrete structural walls for nonlinear seismic analysis. In Proc. of the European Conf. on Struc. Dyn., EURODYN '90, Bochum, Germany, A.A. Balkema, 1991, pp. 471-78.

2. Kabeyasawa, T., Shiohara, H., Otani, S. and Aoyama, H., Analysis of the full-scale seven-story reinforced concrete test structure. Journ. of the Fac. of Eng., Univ. of Tokyo, 1983, 37, pp. 431-78.

3. Vulcano, A., Bertero, V.V. and Colloti, V., Analytical modeling of R/C structural walls. Proc. 9th WCEE, Tokyo - Kyoto, Maruzen 1989, Vol. 6, pp. 41-46.

4. U.S. - Japan Cooperative Research Programs, Test of reinforced concrete structures, Proc. 8th WCEE, San Francisco, Prentice Hall, 1984, Vol. 6, pp. 593 - 706.

5. Damjanić, F.B., Finite element modelling in structural reinforced concrete analysis. In Nonlinear Seismic Analysis of Reinforced Concrete Buildings, eds. P. Fajfar and H. Krawinkler, Elsevier, 1991.

6. Fajfar, P. and Fischinger, M., Nonlinear seismic analysis of RC buildings. Implications of a case study. European Earthquake Engineering, 1987, 1, pp. 31-43.

7. Fischinger, M., Vidic, T. and Fajfar, P., Evaluation of the inelastic response of a R.C. building with a structural wall designed according to Eurocode 8. In Proc. of Int. Conf. Buildings with Load Bearing Concrete Walls in Seismic Zones, Paris, 1991, pp. 487-98.

8. Lybas, J.M. and Sozen, M.A., Effect of beam strength and stiffness on dynamic behaviour of reinforced concrete coupled shear walls. Report No. SRS 444, University of Illinois at Urbana - Champaign, 1977.

9. Keshavarzian, M. and Schnobrich, W.C., Computed nonlinear seismic response of R/C wall-frame structures. Report No. SRS 515, University of Illinois at Urbana - Champaign, 1984.

A Section Analysis Model For The Nonlinear Behaviour Of RC Members Under Cyclic Loading

A.S.ELNASHAI AND K.PILAKOUTAS[1]

Department of Civil Engineering, Imperial College, London SW7 2BU

ABSTRACT

This paper briefly describes a simple program for the nonlinear analysis of reinforced concrete members under cyclic loading based on the section analysis method. Realistic constitutive models are used for steel and concrete, taking into account the effect of cyclic loading. The program is used to investigate the energy dissipation of RC walls and comparisons are made with the results of an experimental programme conducted by the authors.

INTRODUCTION

Under earthquake loading conditions the flexural component of resistance and deformation in RC members provides the principal source of energy dissipation. Capacity design for flexure is simple, and the observed behaviour follows closely the design assumptions. On the other hand, the nonlinear behaviour under reversed loading is complicated and the computational effort required by advanced finite element analysis is considerable. Hence, simple analysis tools based on section analysis methods can provide an economic alternative for investigations on the member level. The assumptions on which the section analysis method is based are as follows:

a) Plane sections remain plane.
b) Full strain compatibility exists between concrete and steel reinforcement.
c) Flexural deformations are independent of shear deformations.
d) The total deformation is the sum of shear and flexural components.

The method of section analysis assumes a certain strain distribution within a section, which is varied until force equilibrium is established. In order to achieve this, material models are used to calculate stresses for given strains. The cyclic models used for steel and concrete are discussed in the following.

---

[1]   Currently, Department of Civil and Structural Engineering, University of Sheffield , Mappin Street, Sheffield, S1 3JD

MATERIAL MODELS

Steel model

The most commonly used steel constitutive relationship in design is the linear elastic perfectly plastic model. There are two significant draw-backs in simplifying the steel behaviour to an elastoplastic model:

a) Ignoring strain hardening.
b) Ignoring stiffness degradation after inelastic load reversals.

The extent of strain hardening varies considerably in different steels; modern codes have identified this and recommend limits for the ratio of ultimate to yield stress. The effect of ignoring strain hardening is to lead to an under-estimate of the section strength. Consequently, the accuracy in estimating the flexural strength, required for the shear design, may be compromised.

The degradation of stiffness under cyclic loading exceeding yield stresses is a material property that cannot be estimated from standard monotonic testing. The degree of degradation has significant impact on the energy dissipated by the steel, hence by the structural member. Additionally, the stiffness of steel on reloading will determine the load level at which concrete will be re-mobilised following the load reversal. Consequently, this influences the dynamic behaviour of the reinforced concrete structure, and should be taken into consideration.

In order to represent the behaviour of the variety of steels used, a tri-linear cyclic model was formulated as shown in figure 1. Monotonic experiments can be used to obtain the model envelope. A yield level is determined by using the monotonic yield load and the strain hardening stiffness 'E_{s1}'. A maximum stress 'f_{su}' is not to be exceeded at any value of strain. Exceeding an ultimate strain 'ε_{su}' will result in the bar fracture and total loss of strength. Loading and unloading up to the yield level and down to zero follows the initial stiffness 'E_{so}'. On reloading, a stiffness 'E_{sa}' is used. Once the yield level is achieved, stress increases according to stiffness 'E_{s1}'.

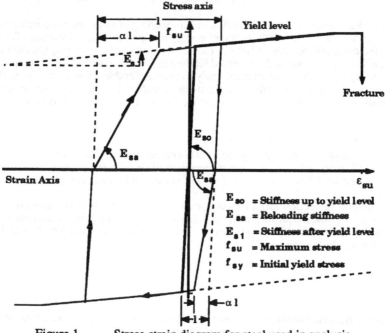

E_{so}	= Stiffness up to yield level
E_{sa}	= Reloading stiffness
E_{s1}	= Stiffness after yield level
f_{su}	= Maximum stress
f_{sy}	= Initial yield stress

Figure 1   Stress-strain diagram for steel used in analysis

The above simple model was developed by using the Massing assumption [1] in a slightly modified form and the stiffness degradation factor 'α' from the work of Santhanam [2]. However, the latter reference used mild steel and as a result the monotonic model includes a yield plateau prior to the onset of strain hardening. This plateau is not observed in most tests on high tensile steel. Consequently, this plateau was removed from the present model for all bar diameters for the sake of simplicity. In the absence of cyclic data for the steels used in the experiments, the value of 'α' is obtained by calibrating the results as discussed later. The implementation of the steel model in a computer program is simple and only the current and previous maximum and minimum permanent strains are required to establish the stress from the current strain.

Concrete model
The concrete model used is based on the work of Mander et al [3,4]. It was chosen for its direct applicability to the method of sections and was implemented with some modifications. The model though uniaxial in its formulation, takes into account the confining stresses.

Concrete confinement :   Experimental work on confinement of concrete was carried out by many researchers and there is a large number of biaxial and triaxial concrete models. However, in most of these experiments active confining stress was applied as a uniform hydrostatic pressure. The effect of steel reinforcement in providing passive confinement has long been recognised and several models have been presented based on RC column experiments. A description of early work is given by Vallenas, Bertero and Popov [5].

The development of the lateral force in the steel depends on the elastic properties of concrete as well as the axial strain. In order to avoid the elaborate calculations for evaluating the confinement factors, the effectiveness of the confinement is provided to the section analysis program as an input. Simple calculations to account for the effect of hoop pattern and spacing, the maximum enhanced confined stress f_{cc}, the strain at which f_{cc} is achieved and ultimate crushing strain ε_{cc}, are given by Tassios [6] in a background document to Eurocode 8.

Monotonic concrete model :   For a uniform confining stress, the monotonic confined compressive stress-strain curve is as shown in figure 2 and is given by:

$$f_c = \frac{f_{cc} \; x \; r}{r - 1 + x^r}$$

(1)

where

$$x = \frac{\varepsilon_c}{\varepsilon_{cc}} \qquad r = \frac{E_c}{E_c - E_{sec}} \qquad E_c = 5{,}000\sqrt{f_{co}} \; \text{(MPa)} \qquad E_{sec} = \frac{f_{cc}}{\varepsilon_{cc}}$$

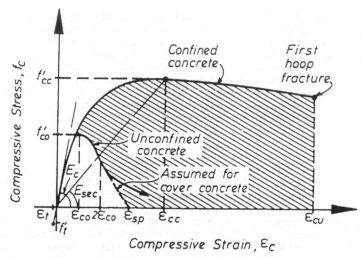

Figure 2   Stress strain model for monotonic loading unconfined and confined concrete [3]

For the unconfined areas, such as the cover and the web area of RC walls, the descending branch after '$2\varepsilon_{co}$' (concrete compressive strain in the longitudinal direction at peak unconfined stress) is defined as a straight line which reaches zero stress at the strain of 'ε_{sp}'. In the computer program implementation, concrete was considered to crush instantaneously, and hence, there is no unloading branch.

For monotonic tensile loading, concrete can carry tensile stresses up to a limit of 'f_t' with stiffness equivalent to the modulus of elasticity 'E_c'. The tensile strength may be affected by micro-cracking initially, and will be lost at the initial stages of cyclic loading. Therefore, in the implementation of the above model the tensile strength was ignored.

<u>Concrete model for cyclic loading</u> :   It is generally accepted that the monotonic stress-strain curve can be considered to be the envelope to the cyclic loading response. To describe a cyclic model, the unloading and reloading branches should be defined as given below.

On complete unloading, the stress drops to zero at a plastic strain 'ε_{pl}'. The value of the plastic strain depends on the previous maximum reversal point (ε_{un}, f_{un}). For strains less than 'ε_{pl}', the stress is considered to be zero. The unloading path is of similar form to the monotonic loading of equation 1, as shown in figure 2, but is modified to pass through the point (0, ε_{pl}), as shown in figure 3. The value of the initial unloading modulus of elasticity 'E_u' has been calibrated by Mander, Priestly and Park [7] on experimental results.

Reloading can occur from different previous unloading history cases and hence the different possibilities ought to be considered, as follows:

a)   Reloading from a strain less than ε_{pl}.
b)   Reloading from a strain 'ε_{ro}' and stress 'f_{ro}' where 'ε_{ro}' is less than 'ε_{un}' and greater than 'ε_{pl}'.
c)   Reloading to a strain higher than 'ε_{un}'.
d)   Reloading to a strain higher than 'ε_{re}'.

For case (a) the stress remains zero until ε_{pl} is exceeded and then case (b) is used to obtain the stresses. A linear stress-strain relationship is assumed between the return point

$(\varepsilon_{ro}, f_{ro})$ and a new target point $(\varepsilon_{un}, f_{new})$ which accounts for the cyclic degradation in strength as shown in figure 3.

$$f_c = f_{ro} + E_r (\varepsilon_c - \varepsilon_{ro})$$ (2)

Where

$$E_r = \frac{f_{ro} - f_{new}}{\varepsilon_{ro} - \varepsilon_{un}}$$

$$f_{new} = 0.92\, f_{un} + 0.08\, f_{ro}$$ (3)

For a strain exceeding 'ε_{un}' and up to 'ε_{re}', Mander et al (1988a) suggested a parabolic transition curve between the target point $(\varepsilon_{un}, f_{new})$ and a common return point $(\varepsilon_{re}, f_{re})$ as shown in equation 4 below.

$$\varepsilon_{re} = \varepsilon_{un} + \frac{f_{un} - f_{new}}{E_r \left(2 + \dfrac{f_{cc}}{f_{co}} \right)}$$ (4)

The value of 'ε_{re}' as obtained from equation 4 is evaluated by using a stiffness higher than the reloading linear branch E_r, since the value of $(2 + f_{cc}/f_{co})$ in the denominator is at least three. This does not conform to observations from a number of cyclic experiments conducted on control cylinders. For the purposes of this work the value of 'ε_{re}' was obtained by dividing 'E_r' by the parenthesis value (instead of multiplying it) and a modified parabolic transition curve has been derived on the same principles. To derive this second order equation, the three boundary conditions used were the two end points $(\varepsilon_{un}, f_{new})$ & $(\varepsilon_{re}, f_{re})$ and the gradient at the first point being E_r. The modified transition curve is given by:

$$f_c = f_{re} + E_{re}\, x + A\, x^2$$ (5)

where $\quad x = (\varepsilon_c - \varepsilon_{re})$

$$A = \frac{(f_{new} - f_{re}) - E_{re} (\varepsilon_{un} - \varepsilon_{re})}{(\varepsilon_{un} - \varepsilon_{re})^2}$$

For strains higher than 'ε_{re}' reloading continues on the monotonic curve as shown is figure 3.

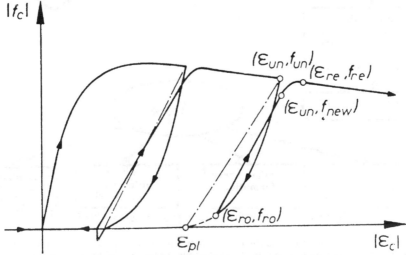

Figure 3   Stress-strain curves for reloading cases [3]

COMPUTER PROGRAM CRELIC

A computer program CRELIC (Cyclic REinforced concrete anaLysis Imperial College) was developed on the basis of section analysis and the material models described above [8]. The solution procedure is outlined by the flow chart in figure 4.

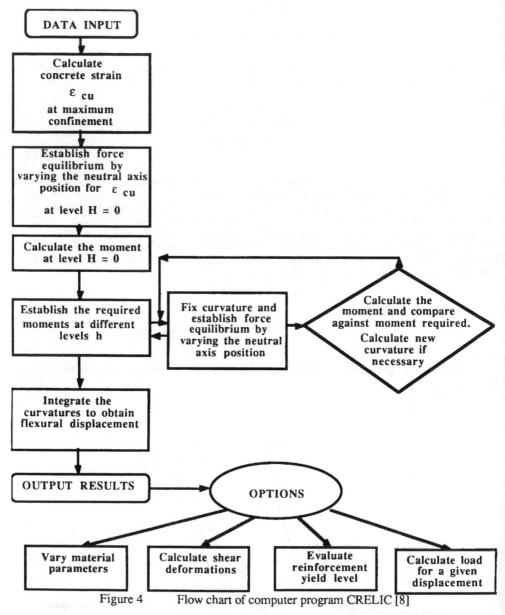

Figure 4   Flow chart of computer program CRELIC [8]

Data input include information on geometry, loading and material characteristics. Control data are also required, to give desired accuracy for the different iterations.

The program calculates automatically the displacement corresponding to the maximum confined strain in concrete. Following that, control is returned to the user through various options. For cyclic loading, the displacement control option enables following the hysteretic behaviour of the member at prescribed displacements or automatic cycling.

Comparisons with a reinforced concrete isolated cantilever wall (SW9)

The program was checked by using the results from the experimental programme during which the work presented in this paper was undertaken [8]. The experimental flexural deformations, obtained by an improved method suggested in ref [8] by integrating the section rotations at various heights, were imposed at the top of the wall. Only a small portion of the output information is presented here in graphical form, and is included in Appendix A (The top horizontal flexural displacement δ3, the vertical displacement δ4 and the displacement at quarter height are shown in figures A1 - A3). The locations of the key displacements selected for output and details of the tested wall are shown in figure 5 below.

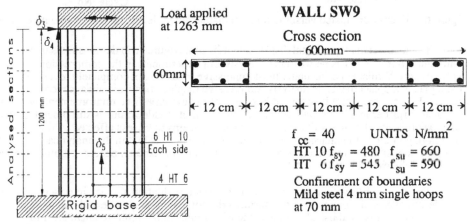

Figure 5   Details of wall SW9 as analysed by CRELIC

Discussion of program results

The load versus flexural displacements δ3 compare well with the experimental data. A slightly higher load is achieved in the analytical results. This may be attributed to the simplification of the post-yield steel behaviour in the analytical model. The relaxation at peak displacement, observed in the experimental results is not seen in the analytical results since this effect has not been modelled. The reloading stiffness is reproduced accurately, and this indicates that the used value of 0.5 for the steel stiffness degradation factor 'α' is reasonable. Lower values of 'α' gave higher initial reloading stiffness and more energy dissipation, as shown in figure 6. The energy dissipation from the analytical results at 'α' equal to 0.5, is still higher than the one from the flexural experimental curves. This is mostly due to the higher strength and no relaxation effects in the analytical model. It is however, less than the overall energy dissipated by the experimental model.

Of particular interest is the difference exhibited by the hysteretic energy per cycle in the early loops as shown in figure 6. The analytical results indicate much less hysteretic energy being absorbed, probably due to the absence of tensile strength in the concrete model used.

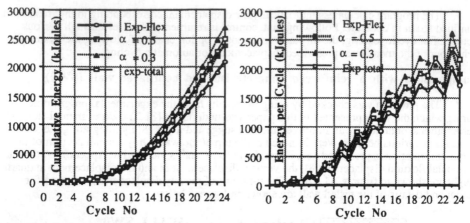

Figure 6   Energy dissipated versus maximum displacement level and corresponding cycle no

The load versus top vertical displacement δ4 plot (figure A.2) demonstrates that a significant difference exists between the analysis and experiments in the compression displacements. Whilst experimentally the compressive loads do not completely reverse the tensile plastic displacements, in analysis compressive displacements are always recovered even after yield. The same effect is observed in strains of the extreme fibre reinforcement, whilst for bars located further inside the wall section, the reversal of tensile strains is still much less than in the experiments. The net wall extension, however, seems to be less affected. A difference worth noting in the wall extensions on reloading (figure A.3) is that in the analytical results, the minimum extension for a particular cycle after significant inelasticity, occurs at a much higher load than in the experiments.

The above observations may be a result of concrete dilation at crack interfaces. Due to shear deformations, the opposing crack surfaces do not match on closure and due to their roughness, contact is established even when tensile strains in the reinforcement are still observed. This mechanism, termed 'aggregate interlock', is considered to be responsible for shear transfer through cracks as a consequence of the co-existent normal stresses. Tassios and Vintzeleou [9] demonstrated experimentally the shear transfer mechanism and concrete dilation of cracked surfaces during cycling for low constant normal stresses. Dilation degradation was also noted to occur during load cycling. An attempt to evaluate the dilation effects on the concrete stress-strain cyclic relationship was made by Zhu, Wu and Zhang [10] who proposed an empirical model based on a limited number of experimental results. Unfortunately, as with many other shear stiffness formulations, explicit use is made of a crack width, which cannot be calculated directly by the method of section analysis.

The results from the flexural analytical model seem to reproduce accurately the strength of the cyclic test, and reasonably accurately the general behaviour of reinforced concrete walls subjected to cyclic loading. The flexural model was utilised further in ref [8], to investigate a wide range of parameters which extended beyond the scope of the experimental investigation.

CONCLUSIONS

(a) The method of section analysis can be used to investigate the nonlinear behaviour of RC members under cyclic loading, provided suitable material models are used.

(b) Strain hardening and stiffness degradation of steel reinforcement during cyclic loading of RC in the inelastic range have a great influence on the section capacity, member stiffness and energy dissipation capacity.

(c) Reloading of RC sections mobilises stresses in concrete even in the presence of tensile strains. This is due to the incomplete closure of cracks and can also be viewed as concrete dilation. This phenomenon is usually neglected in concrete material models. This leads to inaccuracies in the reloading branches for cyclic loading.

ACKNOWLEDGEMENT

The work described above is part of an on-going research project on the seismic behaviour, repair and upgrading of RC members under cyclic and earthquake loading. Funding was provided by the Science and Engineering Research Council and the ESEE Section at Imperial College. The experimental work was undertaken in the Concrete Structures laboratory at Imperial College.

Work is continuing at Imperial College on the response of RC walls with low shear ratios subjected to high cyclic shear, analytical local/global models and selective repair schemes for earhquake - damaged RC members. Complimentary work is being planned at the University of Sheffield.

REFERENCES

1. Hays,C.O., "Inelastic Material Models in Earthquake Response", Journal of Structural Engineering, ASCE, Vol. 107, No. 1, January 1981, pp 13-28

2. Santhanam, T.K., "Model for Mild Steel in Inelastic Frame Analysis", Journal of Structural Engineering, ASCE, Vol. 105, No. 1, January 1979, pp 199-220

3. Mander, J.B, Priestley, M.J.N. & Park, R., "Theoretical Stress-Strain Model for Confined Concrete", Journal of Structural Engineering, ASCE, Vol. 114, No. 8, August 1988a, pp 1804-1826

4. Mander, J.B, Priestley, M.J.N. & Park R., "Observed Stress-Strain Behaviour of Confined Concrete", Journal of Structural Engineering, ASCE, Vol. 114, No. 8, August 1988b, pp 1827-1849

5. Vallenas, J.M., Bertero, V.V. and Popov, E.P., "Concrete confined by rectangular hoops and subjected to axial loads", Report No EERC-77/13, Earthquake Engineering Research Centre, University of California, Berkeley, August 1977

6. Tassios, T.P., "Specific rules for concrete structures", In Background Document for Eurocode 8 - Part 1, Volume 2 - Design Rules, Commission of European Communities, 1989, pp 1-123

7. Mander, J.B, Priestley, M.J.N. & Park, R., "Seismic design of bridge piers", Research Report No. 84-2, University of Canterbury, New Zealand, 1984

8. Pilakoutas K., "Earthquake resistant design of reinforced concrete walls", PhD Thesis, University of London, 1990

9. Tassios, T.P. and Vintzeleou E., "Concrete to Concrete Friction", Journal of Structural Engineering, ASCE, Vol. 113, No. 4, April 1987, pp 832-849

10. Zhu, B., Wu, M. and Zhang, K., "A Study of Hysteretic Curve of Reinforced Concrete members under Cyclic Loading", 7th World Conference in Earthquake Engineering, Vol. 6, Istanbul, Turkey, 1980, pp 509-516

APPENDIX A

Figures of Force versus Displacement for comparison between analysis and experiments

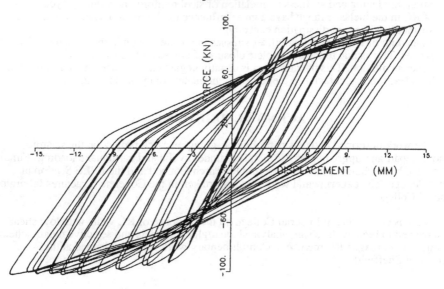

ANALYSIS

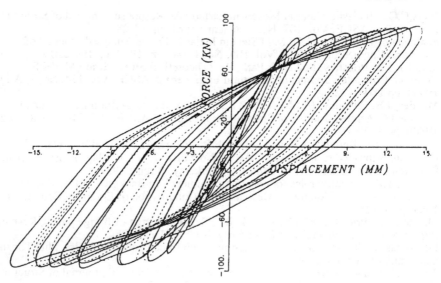

EXPERIMENT

Figure A-1. The top horizontal displacement δ_3

213

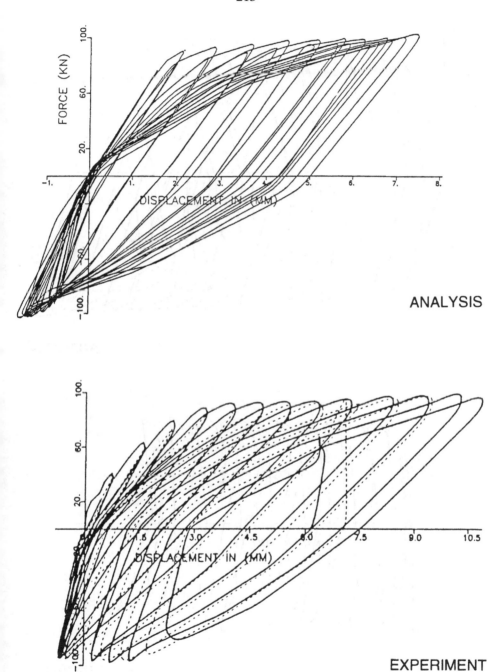

ANALYSIS

EXPERIMENT

Figure A-2. The vertical displacement δ_4

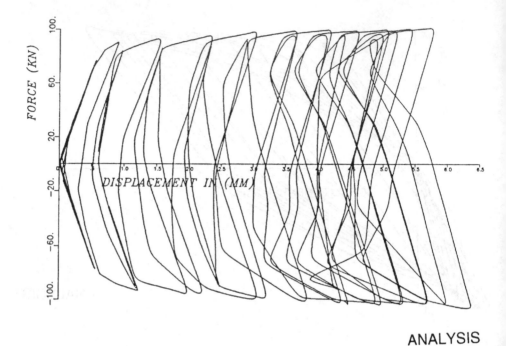

ANALYSIS

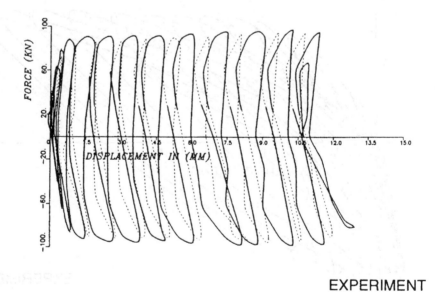

EXPERIMENT

Figure A-3. The vertical displacement at quarter height

TWO- AND THREE-DIMENSIONAL NONLINEAR FINITE-ELEMENT ANALYSIS OF STRUCTURAL WALLS

MICHAEL D. KOTSOVOS
Department of Civil Engineering, National Technical University of Athens,
42 Patission Street, Athens, Greece.
MILIJA N. PAVLOVIĆ
Department of Civil Engineering, Imperial College, London SW7 2BU, U.K.
IOANNIS D. LEFAS
Balfour Beatty Projects and Engineering Ltd.,
Marlowe House, Sidcup, Kent DA15 7AU, U.K.

ABSTRACT

This article attempts to elucidate the relative merits of two- and three-dimensional finite-element versions of a brittle material model for structural concrete when applied to the analysis of structural walls. It is found that, while both versions yield satisfactory results for strength predictions and cracking processes, the 3-D option appears to lead to considerable improvement of deformation characteristics.

INTRODUCTION

The difficulties encountered in the search of a general, reliable nonlinear finite-element (FE) model for structural concrete are well known. The proliferation of many different models in recent years is further complicated by their constant "evolution" - a direct result of their usual lack of generality and reliability, since the reported good agreement between prediction and experiment obtained by "tuning" parameters with respect to a given structural member often breaks down as soon as a different member and/or structural type are tackled. Moreover, the number and complexity of the parameters needed as input to such models bear little relation with the sort of material data readily available to designers.

Over the last decade, the development of a simple, reliable FE model of general applicability has been carried out at Imperial College (IC), a task based on earlier fundamental research of concrete at the material level. What has emerged is that the difficulties of many of the other models proposed to date stem from the introduction of many parameters which either are of secondary importance (e.g. "aggregate interlock", "dowel action", "bond slip", "tension stiffening", etc.) or are simply non-existent in reality (this is especially true of the so-called "strain softening"). Conversely, the lack of recognition of the importance of triaxial effects - invariably present in any structural component prior to collapse and governing the mechanism of the latter - also explains why much of the existing FE modelling of concrete structures is unrealistic. In contrast to these shortcomings, the model developed at IC describes the constitutive relations and failure criteria of the material under any triaxial stress state, such laws having been obtained under definable experimental conditions and requiring only the most basic of material parameters - namely the cylinder strength in uniaxial compression (f_c) - for their description.

The proposed FE model has been fully described in earlier publications which first dealt with two-dimensional (2-D) problems [1-4], and subsequently extended the numerical scheme to three-dimensional (3-D) structures [5-8]. While these publications were concerned with general outlines of the model itself and its application to a wide range of structural forms, interest has also been focussed on the numerical program's potential for studying specific structural types and/or problems, as regards, for example, the elucidation of the actual mechanism of "punching" failure in slabs [9] or the effect of concrete strength in beams [10]. The present article addresses one such specific problem, namely the mathematical simulation of structural walls on the basis of the brittle model in question. Emphasis is placed on the numerical modelling of this structural type by means of both 2-D and 3-D options, the structural response up to failure and the collapse mechanism at ultimate load, and the ensuing recommendations for both FE discretization and design of the components themselves. The analytical predictions and various observations are backed by relevant experimental data obtained at IC [11].

TWO-DIMENSIONAL MODELLING

The modelling of most reinforced-concrete (RC) structures which can be reduced to a 2-D problem may be achieved by means of an 8-node isoparametric element for concrete and a 3-node bar element for the reinforcing steel [1-4]. Both elements are well tested and widely used, and since they deform following the same pattern, compatibility of deformation between steel and concrete is ensured. The only prerequisite which has to be followed is that all the 3-node bar elements must lie on the boundaries of existing 8-node isoparametric elements. This implies that the generation of the steel mesh should follow the generation of the concrete mesh.

In mimicking the behaviour of RC structural walls subjected to loads acting in their own plane, the 2-D formulation based on the assumption that plane-stress conditions apply is an obvious and attractive procedure on account of both simplicity and economy. It is, however, important to assess the accuracy of such an approach, and any inherent shortcomings which only the more formal 3-D modelling can remove.

That the 2-D version gives good predictions of load-carrying capacity for a wide range of structural-wall tests reported by various workers (encompassing both horizontally-loaded components [12,13], as well as members subjected to combined vertical and horizontal loads [14-16]), may be seen by reference to Fig. 1 [17]. The analytical results were obtained by adopting a constant 7*7 mesh of concrete elements with steel-bar elements lumped at discrete locations along the boundaries of the grid (Fig. 2). Therefore, even if the original structure had a denser arrangement of reinforcement than the mesh used, the equivalent area of the flexural reinforcement and of the confining steel in the edge members was placed at the boundaries of the two extreme columns of concrete elements, while the appropriate area of steel corresponding to the horizontal and vertical web reinforcement was arranged along the boundaries of the intermediate concrete elements. Base fixity conditions were achieved by simply fixing the degrees of freedom of the relevant nodes in the mesh.

In all experiments, a top beam or a top slab acted as the distributor of the applied vertical and horizontal loading. The necessary edge-pressure load type was selected from the load library, so as to apply a uniformly distributed traction and/or pressure to the series of the top elements. In the particular case of the walls, constant pressure simulated the constant applied vertical load (and/or the weight of the heavy top slab or beam), while edge traction simulated the incrementally applied horizontal load on the wall up to failure. The choice of the horizontal load-step size was based on the accuracy required to predict the maximum load-carrying capacity of the structural wall and was typically of the order of 5%-10% of the estimated strength.

Prior to undertaking the numerical analysis, the relevant experimental information regarding the various parameters is obviously necessary. This includes the geometry of the wall, the location and amount of the reinforcement, the material properties (uniaxial compressive cylinder strength for the concrete (f_c), and yield strength (f_y) and other

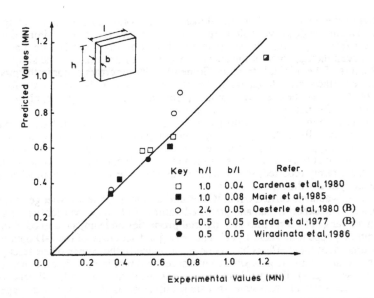

Figure 1. Correlation of predicted load-carrying capacity of walls with experimental values.

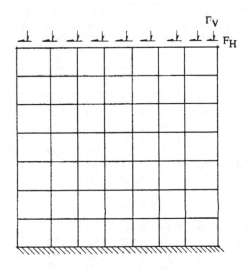

Figure 2. Typical finite element mesh adopted for the 2-D analysis of walls.

characteristics of the steel), as well as actual loading and boundary conditions. Despite the fact that all these data are seldom available in full detail in the references containing other authors' tests, it is clear from Fig. 1 that the errors incurred in adopting the 2-D model are within the order of 10%, even though some of the analyses were based on considerable simplifications to the actual features of the original structure.(Somewhat larger errors occur in the predictions of barbell specimens (these are referenced by the letter B); although the worst error for the latter - of the order of 20% - may be deemed to be acceptable for practical purposes, higher accuracy is achievable through the more exact 3-D modelling.)

The overall trend of the deformational properties of structural walls may also be estimated by 2-D modelling although the high accuracy of the ultimate-load predictions is not usually achieved. A typical example may be seen in Fig. 3. One reason for the stiffer response of the numerical modelling is due to time-dependent effects (in the course of the loading history) which are rarely reported, so that no allowance can be made for them when attempting to reproduce analytically the tests of others.

Structural-wall tests carried out at IC ensured a more precise knowledge of the various input parameters. The experimental and predicted values of ultimate load for twelve such walls (encompassing a range of geometries and material properties and subjected to combined loading conditions) appear in Table 1 [11]. The very good accuracy of the 2-D option is evident. The time-dependent effects having been estimated in the course of the tests [11], enables now a reduction in the gap between measured and predicted deformational responses to be achieved (see Fig. 4). This gap between stiffnesses, though halved, still persists, which implies that the fully-triaxial stress states that are set up close to the ultimate load appear to affect more the deformation rather than the strength characteristics of walls; this is particularly

TABLE 1
Data for walls tested and comparison of experimental and predicted ultimate horizontal loads.

WALL	CUBE STRENGTH (MPa)	REINFORCEMENT PERCENTAGE				AXIAL LOAD (kN)	ULTIMATE HORIZ. LOAD (kN)		ANAL/ EXPER
		hor.	vert.	flex.	confin.		exper.	anal. 2-D/3-D	
				Type I (h/l =1)					
SW11	52.3	1.10	2.40	3.10	1.20	0	260	243/245.7	0.94/0.95
SW12	53.6	1.10	2.40	3.10	1.20	230	340	304/300.3	0.89/0.88
SW13	40.6	1.10	2.40	3.10	1.20	355	330	272/286.7	0.82/0.87
SW14	42.1	1.10	2.40	3.10	1.20	0	265	216/232.1	0.82/0.88
SW15	43.3	1.10	2.40	3.10	1.20	185	320	257/273.0	0.80/0.85
SW16	51.7	1.10	2.40	3.10	1.20	460	355	320/341.3	0.90/0.96
SW17	48.3	0.37	2.40	3.10	1.20	0	247	202/204.8	0.82/0.83
				Type II (h/l =2)					
SW21	42.8	0.80	2.50	3.30	0.90	0	127	124.8/115.4	0.98/0.91
SW22	50.6	0.80	2.50	3.30	0.90	182	150	146.3/143.0	0.98/0.95
SW23	47.8	0.80	2.50	3.30	0.90	343	180	146.3/143.0	0.81/0.79
SW24	48.3	0.80	2.50	3.30	0.90	0	120	127.9/123.6	1.07/1.03
SW25*	45.0	0.80	2.50	3.30	0.90	325	150	146.3/143.0	------/-----
SW26	30.1	0.40	2.50	3.30	0.90	0	123	96.0/ 90.6	0.78/0.74

* = premature exper.failure
TENSILE STRENGTH f_{su} (MPa)
Flex. reinforcement of edge element and Vert. web reinforcement = 560
Horiz. web reinforcement = 610
Confinement reinforcement = 490

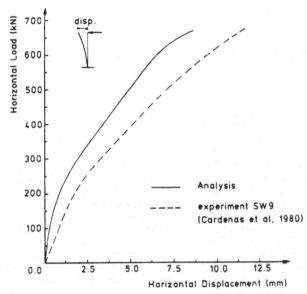

Figure 3. Typical load-displacement curve of wall (2-D analysis).

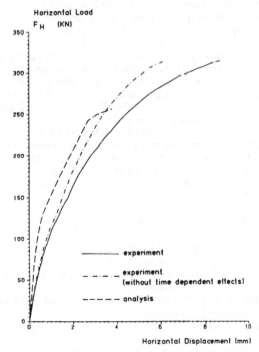

Figure 4. Experimental and analytical horizontal load-top horizontal displacement curves for wall SW15 (2-D analysis).

true for low height-to-width ratios and for the more heavily reinforced walls in which the high percentage of main reinforcement precludes the more readily predictable flexural type of failure.

The above satisfactory correlations between analysis and experiment of structural walls (or that part of the wall representing the critical storey element of a structural wall system) extend not only to strength and deformation characteristics, but also to crack patterns and modes of failure. This has enabled a reasonably comprehensive parametric study of structural-wall behaviour to be undertaken on the basis of the 2-D model [17]. The whole investigation was complemented by conducting further select experiments at IC [18,19]. A comparative study of the results of this experimental work and of the predictions obtained by using the 2-D FE package is reported elsewhere [20]. Several conclusions may be drawn on the basis of the above work. The first of these concerns the causes and mechanism of failure, which may be illustrated with the help of Fig. 5, where the crack pattern and deformed shape of a typical wall are given for various load levels. Thus Fig. 5a shows how the initiation of flexural cracking occurs at an early stage, corresponding to quite low load levels. The early inclined cracking is depicted in Fig. 5b, while Fig. 5c illustrates a stage of extensive flexural and inclined cracking. Clearly, macrocracking initiates in the lower tensile zone of the wall, thereafter spreading progressively towards the lower compressive zone as the load increases. In the load step just prior to failure, the depth of the compressive zone has been reduced to only a small portion of the wall width (Fig. 5c). The mode of failure is characterized by longitudinal cracking within the compressive zone near the base of the wall, as indicated in Fig. 5d. In this region of maximum flexural action, the biaxial compressive stress state is in excess of the uniaxial-strength value, but, as always, failure is triggered by compression-tension stresses in immediately-adjacent locations [21,22].

Another conclusion stemming from the above failure mechanism is that the often-invoked truss analogy does not form a rational basis for the design of RC structural walls. It is evident that the "web" region contributes insignificantly to the strength of the wall and that it is the compressive zone of the "flange" which governs the ultimate-load capacity. Such behaviour is fully compatible with the compressive-force path concept and associated notions of unavoidable triaxiality leading, invariably, to "tensile"- never "compressive"- failures [21,22]. Wall capacity , therefore, is enhanced by strengthening the compressive zone (e.g. by concentrating the vertical reinforcement at the wall edges and/or adding confining reinforcement to enhance triaxiality there), allowing, at the same time, the reduction of web reinforcement (both vertical and horizontal) to near-nominal levels. The presence of axial load is also beneficial as it increases the depth of the compressive zone.

Finally, neither the effect of height-to-width ratio (h/l or, in the case of storey elements, M/Vl) nor the amount of vertical reinforcement are properly accounted for in current code provisions for structural walls. Nor do these distinguish between under-reinforced or over-reinforced cases, with the result that ensuing estimates of flexural capacity might turn out to be inadequate. Furthermore, certain code provisions may be shown to be unsafe for low-to-medium concrete strengths [17].

THREE-DIMENSIONAL MODELLING

As the 3-D model has been described elsewhere [5-8], only its main features need to be listed presently before concentrating on its application to structural walls. The model uses a quantitative formulation, based on fully-triaxial test data of the stress-strain relationships and of the failure envelope of concrete which are uniquely defined upon specification of the cylinder strength f_c. The "constant" parameters for general use of the model were established by a previous, extensive objectivity study [6]. As a result

(i) the "shear-retention factor" is set to 0.1 on the basis of both physical reasoning and numerical grounds;

(ii) the "single crack" approach is related to the requirement of gradual redistribution of residual forces due to cracking and the minimization of the risk of numerical instability due to explosive cracking;

(iii) the immediate update of the D-matrices of the newly-cracked Gauss points is implemented so as to reduce the risk of propagation of spurious mechanisms;

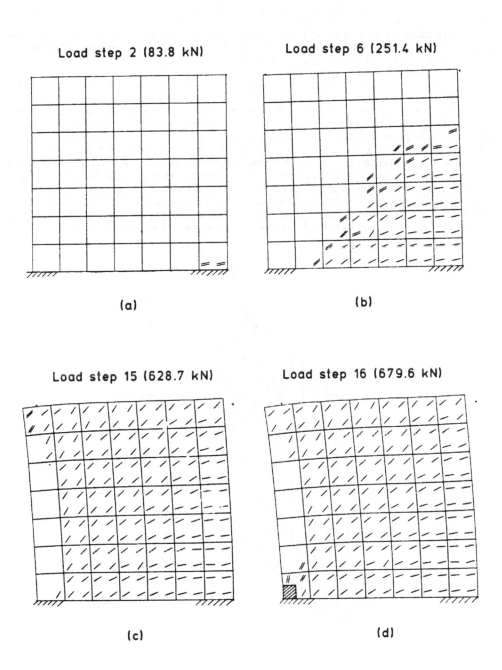

Figure 5. Crack pattern and deformed shape of typical wall at various load stages. (2-D analysis)

(iv) "under-integration" of the quadratic (serendipity, but not Lagrangian) solid elements is seen as a generally desirable feature (in agreement with linear analysis, but against the claims of some analysts in the field of nonlinear FE modelling of structural concrete).

All cases reported in Table 1 were rerun by using 3-D modelling. Figure 6 shows the FE discretization adopted for the analysis. The mesh has 25 HX20A elements (228 nodes and 684 degrees of freedom) for the concrete, and 75 bar elements for the steel reinforcement. The total area of vertical reinforcement, and the area of vertical reinforcement held by the stirrups, are equal to those of the tests. Allowance was made for the presence of the two concealed columns in the wall, which resulted in a denser vertical steel distribution at the edges where the compressive force is critical. However, the distribution deviates from the uniform one in the tests as the actual steel areas encompassed by each finite element have been lumped onto its edges. As regards the horizontal reinforcement, the area was adjusted so as to keep the same area per unit vertical length (the spacing in the test was 80mm compared to 150mm in the analysis). The relevance of smearing of the reinforcement will be discussed below. Another deviation from the test conditions relates to the boundary loading. In the test, the load was applied on one of the lateral faces of the top beam/flange referred to earlier, whereas the present analysis applies the load by means of surface tractions distributed uniformly on the five top finite elements of Fig. 6. Nevertheless, the latter deviation from the test loading appears to play a secondary role, although its true relevance will be assessed in forthcoming work.

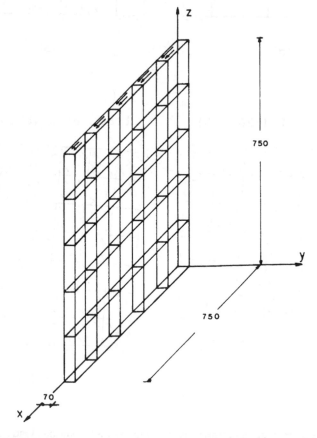

Figure 6. Typical 3-D finite element mesh of the wall , consisting of 25 brick elements.

Table 1 summarizes the ultimate horizontal analytical load at divergence. Figures 7 and 8 show the analytical crack patterns (and displacements magnified by a factor of 10) up to failure of a typical wall, and a comparison of the analytical and experimental load-deflection curves for a suite of walls, respectively. The first flexural cracks appear in the tensile zone of the built-in edge at a total applied load of 68.3kN (see Fig. 7). Subsequently, additional flexural cracks occur within the tensile zone of the wall and, then, cracking propagates extensively throughout the diagonal of the wall. The splitting of the compressive zone - accompanied by the yielding of the tension steel - at the built-in support initiates at 218.4kN and leads to the collapse of the wall (i.e. at the load step following maximum sustained load), which in the present analysis takes place for a total applied load equal to 245.7kN. Such a failure mechanism compares very well with the experimental mode of failure, as attested by the experimental crack pattern and corner splitting described in Ref. 11. However, the analytical failure load underestimates the experimental value by 10% (5% should the load at divergence be considered). In this respect, it appears that the adopted FE discretization may not model adequately the confinement introduced by the stirrups in the compressive zone at the corner, since the spacing is about twice that of the test. Finally, Fig. 8 shows that the analytical load-deflection curves are stiffer than their experimental counterparts. (This figure includes the results of both 2-D and 3-D analyses.) Clearly, in structural-wall tests, experimental displacements involving early stages of long-term effects become significant due to the extensive diagonal cracking of the member and the yielding of the tension reinforcement.

DISCUSSION AND CONCLUSIONS

By reference to Table 1, in which the experimental failure loads are compared with the results of 2-D (7*7 mesh) and 3-D (5*5*1 mesh) analyses, it may be seen that ultimate-load predictions (all of which, incidentally, refer to the load at which numerical divergence occurs) are generally reasonably accurate irrespective of the type of analysis used. However, it is evident that even a coarser 3-D mesh usually leads to improvements over its more refined 2-D counterpart. The exception to this observation appear to be slender walls with no axial (i.e. vertical) load and low levels of concrete strength (i.e. walls SW21 and SW26). For these wall tests, two additional 3-D runs were carried out with a finer mesh (7*7*1): the results gave an improvement for SW26 (98.8 kN, with the corresponding analytical to experimental load ratio of 0.8) but no change from the 5*5*1 mesh result in the case of SW21.

The advantages of adopting a 3-D model are particularly noticeable for "squat" walls (h/l=1) since the high lateral applied forces lead to large stresses in the region of the lower compressive zone, which is best described through 3-D modelling. Conversely, in the case of "slender" walls (h/l=2), the stress field in this critical lower compressive region has a lower intensity, resembling the plane-stress assumptions, so that the 2-D model gives adequate predictions provided that the reinforcement distribution has been sensibly described.

As regards the effect of concrete strength, it may be said that, as this material parameter increases, the gap between 2-D and 3-D predictions narrows since, as is well known [10], the degree of triaxiality becomes less important for larger values of f_c. Therefore, there is little to be gained by using the more formal 3-D model when the cube strength is above a certain value (say, 45 MPa). This does not imply that triaxiality is unimportant beyond this value (in fact, one must go to much higher values before attaining the "high-strength" concrete range where triaxiality plays a markedly less decisive role); rather, it would appear that predictions beyond $f_c \sim 40$ MPa are insensitive to the type of model adopted.

Implicit in the above observations is the notion that, when triaxial effects are important, attention should be paid to the reinforcement detailing in the critical zones. This is particularly true as regards the role of stirrups as a means of achieving suitable confinement [10,23].

Finally, whereas ultimate-load and crack-pattern predictions show that both 2-D and 3-D models provide adequate alternatives in FE representations of structural walls, the analytical results for the deformational levels indicate the clear superiority of the 3-D option over its 2-D counterpart. Once the time-dependent effects in the tests have been allowed for (a mandatory step, in view of the very considerable magnitudes of the ensuing displacements - see Fig. 8), 2-D predictions still tend to underestimate the deformational response. On the other hand, 3-D

224

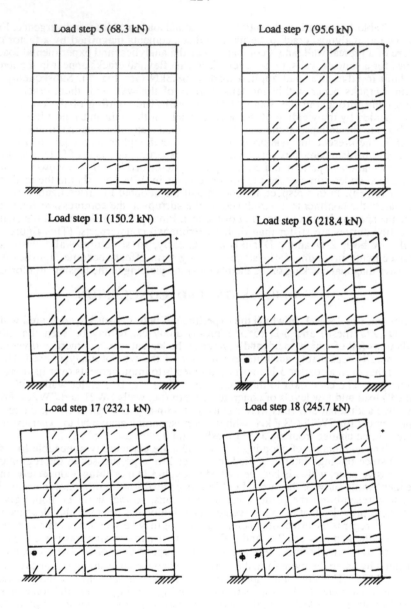

Figure 7. Structural wall SW11. Crack patterns at some load levels up to failure. (3-D analysis)

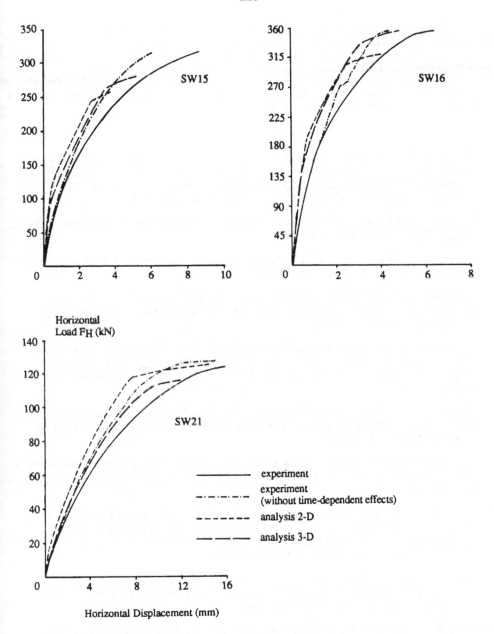

Figure 8. Predicted and experimental horizontal load-top horizontal displacement curve for wall specimens SW15, SW16 and SW21 (2-D and 3-D analyses).

analyses lead to very considerable improvements in displacements, even though the initial stiffness is very similar to that of the 2-D model (the reasons for this are the neglect of cover [6,7] and microcracking stemming from ambient laboratory conditions prior to testing [11]). Such an improvement is particularly evident for low axial loads irrespective of the height-to-length ratio of the wall (see, typically, the plots for SW15 and SW21 in Fig. 8); conversely, for high levels of pre-compression, there is little difference between 2-D and 3-D displacement predictions (see the plot for SW16 in Fig. 8).

REFERENCES

1. Kotsovos, M.D., Pavlović, M.N. and Arnaout, S., Nonlinear finite element analysis of concrete structures: A model based on fundamental material properties. In NUMETA 85. Numerical Methods in Engineering: Theory and Applications, eds. J. Middleton and G.N. Pande, A.A. Balkema, Rotterdam/Boston, 1985, Vol.2, pp. 733-41.

2. Bédard, C. and Kotsovos, M.D., Applications of NLFEA to concrete structures. J. Struct. Engrg., ASCE, 1985, 111, 2691-707.

3. Bédard, C. and Kotsovos, M.D., Fracture processes of concrete for NLFEA methods. J. Struct. Engrg., ASCE, 1986, 112, 573-87.

4. Kotsovos, M.D. and Pavlović, M.N., Non-linear finite element analysis of concrete structures: Basic analysis, phenomenological insight, and design implications. Engrg. Computation, 1986, 3, 243-50.

5. González Vidosa, F., Kotsovos, M.D. and Pavlović, M.N., Three-dimensional finite element analysis of structural concrete. In Computer Aided Analysis and Design of Concrete Structures, eds. N. Bićanić and H. Mang, Pineridge Press, Swansea, 1990, Vol. 2, pp. 1029-40.

6. González Vidosa, F., Kotsovos, M.D. and Pavlović, M.N., A three-dimensional nonlinear finite-element model for structural concrete. Part 1: Main features and objectivity study. Proc. ICE (Part 2), forthcoming.

7. González Vidosa, F., Kotsovos, M.D. and Pavlović, M.N., A three-dimensional nonlinear finite-element model for structural concrete. Part 2: Generality study. Proc. ICE (Part 2), forthcoming.

8. González Vidosa, F., Kotsovos, M.D. and Pavlović, M.N., Nonlinear finite-element analysis of concrete structures: performance of a fully three-dimensional brittle model. In Computational Structures Conference (to be held at Heriot-Watt University, 20-22 August 1991). (Also, Computers & Structures, forthcoming.)

9. González Vidosa, F., Kotsovos, M.D. and Pavlović, M.N., Symmetric punching of reinforced concrete slabs: An analytical investigation based on nonlinear finite-element modelling. ACI Struct. J., 1988, 85, 241-50.

10. Seraj, S.M., Kotsovos, M.D. and Pavlović, M.N., Three-dimensional finite-element modelling of normal-and high-strength reinforced-concrete members, with special reference to T-beams . In Computational Structures Conference (to be held at Heriot-Watt University, 20-22 August 1991). (Also, Computers & Structures, forthcoming.)

11. Lefas, I.D., Behaviour of reinforced concrete walls and its implication for ultimate limit state design. Ph.D. Thesis, Imperial College, University of London, 1988.

12. Cardenas, A.E., Russell, H.G. and Corley, W.G., Strength of low rise structural walls. In Reinforced Concrete Structures Subjected to Wind and Earthquake Forces, SP-63, American Concrete Institute, Detroit, 1980, 221-41.

13. Osterle, R.G., Fiorato, A.E., Aristizabal-Ochoa, J.D. and Corley, W.G., Hysteretic response of reinforced concrete structural walls. In Reinforced Concrete Structures Subjected to Wind and Earthquake Forces, SP-63, American Concrete Institute, Detroit, 1980, 243-73.

14. Barda, F., Hanson, J.M. and Corley, W.G., Shear strength of low-rise walls with boundary elements. In Reinforced Concrete Structures in Seismic Zones, SP-53, American Concrete Institute, Detroit, 1977, 149-202.

15. Maier, J. and Thürlimann, B., Bruchversuche an Stahlbetonscheiben, Institut für Baustatik und Konstruktion, Eidgenössische Technische Hochschule, Zurich, 1985.

16. Wiradinata, S. and Saatcioglou, M., Tests of squat shear walls under lateral load reversals. In Proceedings of the 3rd U.S. National Conference in Earthquake Engineering, Earthquake Engineering Research Institute, 1986, 2, 1395-406.

17. Lefas, I.D. and Kotsovos, M.D., NLFE analysis of RC structural walls and design implications. J. Struct. Engrg., ASCE, 1990, 116, 146-64.

18. Lefas, I.D., Kotsovos, M.D. and Ambraseys, N.N., Behaviour of R.C. structural walls : Strength, deformation characteristics and failure Mechanism. ACI Struct. J., 1990, 87, 23-31.

19. Lefas, I.D. and Kotsovos, M.D., Strength and deformation characteristics of RC Walls under load reversals. ACI Struct. J., 1990, 87, 716-26.

20. Lefas, I.D. and Kotsovos, M.D., Nonlinear finite element analysis of R.C. structural walls. In Proceedings of the 4th International Conference on "Computational Methods and Experimental Measurements, Capri, Italy, Vol. Computers and Experiments in Stress Analysis, 25-34.

21. Kotsovos, M.D., Compressive force path concept: A suitable basis for reinforced concrete ultimate limit-state design. ACI Struct. J., 1988, 85, 68-75.

22. Pavlović, M.N., Structural concrete and ultimate-strength philosophy: Towards a more rational approach to analysis and design. In The Art of Structural Engineering (A Conference in Honour of Professor Emeritus Leonard Kelman Stevens), The University of Melbourne, 1991, 171-95.

23. Lefas, I.D., Ductility evaluation of reinforced concrete structural walls. In Proceedings of the International Conference on Buildings with Load Bearing Concrete Walls in Seismic Zones, Paris, 1991.

FINITE ELEMENT MODELLING IN STRUCTURAL REINFORCED CONCRETE ANALYSIS

FRANO B. DAMJANIC

Institute of Structural and Earthquake Engineering,
Department of Civil Engineering, University of Ljubljana,
Jamova 2, 61000 Ljubljana, Slovenia.

ABSTRACT

A finite element modelling for the static, cyclic and dynamic/seismic analysis of structural reinforced concrete walls is considered. The nonlinear material model briefly described proves its robustness even in a case of the lack of complete material data. Three numerical examples illustrate the applicability of the model and code developed and also provide a comparison between numerical and experimental results.

INTRODUCTION

A large number of numerical models and techniques for the analysis and design of reinforced concrete structures has been developed, and the literature abounds with examples of their implementation and application (See for example Refs. [1,2]). On the other hand, the advent of powerful computers, as well as microcomputers with parallel processing, endowed with exciting prospect of developing an interactive design system, gives the possibility that such techniques can be efficiently employed to the solution of complex practical problems. However, it is true to say that the numerical capability is nowadays in advance of the knowledge of concrete constitutive behaviour. Namely, although reinforced concrete is probably the most commonly used of all structural materials, concrete constitutive properties have not been completely identified, and currently there is no generally accepted material law available to model its behaviour. The complexities of concrete behaviour under different conditions and existing experimental evidence are such that it is therefore extremely difficult, if not impossible, to establish one numerical model which can be adequate for every real situation. Certain idealizations and approximations have to be employed. As a result, a large "family" of models and approaches have been proposed. Among them there is a large number of very sophisticated models whose accuracy far outweighs the degree of accuracy and reliability of the material properties. Some of these properties or physical parameters involved in the models, in some cases, can not be even experimentally determined. Having experience in both the fields, finite element development and practical concrete design, we are aware of all the problems. Therefore, we tried to develop a model which is relatively simple, but capable of providing an accurate enough estimate of structural response for practical engineering design. The constitutive model for concrete employs available basic material data only (e.g. Young's modulus, uniaxial compressive and tensile strength, Poisson ratio, and ultimate compressive strain). In fact, with chosen concrete quality, these properties can be simply estimated from a code of practice. Nevertheless, the model developed accounts for the most dominant nonlinear behaviour of reinforced concrete; tensile cracking, nonlinear multiaxial compressive behaviour of concrete, and yielding of reinforcement.

The purpose of this paper is to illustrate the use of F.E. model for the static, cyclic and dynamic/seismic analysis of R.C. structural walls. The model was developed earlier [3-5] and further improved and extended [6-10] primarily for the analysis of reinforced and prestressed concrete structures subjected to short term static and transient thermal loading. The constitutive model is incorporated into our finite element package BET50, which has been aready successfully applied for the analysis of several practical problems [10-13]. Three types of R.C walls are analysed in this paper. Results obtained illustrate the applicability of the model and the code.

OUTLINE OF MATERIAL MODELLING AND SOLUTION PROCEDURE

Detailed formulations of material modelling and solution procedure have been provided elsewhere (e.g.[4,7]), and only a brief description is outlined herein.

Material modelling

The constitutive model adopted accounts for the most dominant nonlinear behaviour of concrete. Both a perfect and a strain hardening viscoplasticity approach can be employed to model the multiaxial compressive behaviour. The yield criterion is assumed to be dependent on the two first stress invariants, resulting in a Drucker-Prager type loading surface with slightly curved meridians [7]. Either a simple linear hardening rule or one extrapolated from a uniaxial stress-strain test (Fig. 1(a)) can be used to control the motion of the loading surfaces. Concrete crushing is strain controlled with a crushing surface being adopted similar to the yield surface [7]. A linear elastic tensile behaviour is assumed until the cracking surface is reached (tension cut-off criterion). A smeared fixed crack approach is adopted. After the appearance of the first crack at any integration point, a second crack, orthogonal or in the range between 30^0 and 90^0 to the first one, can be noticed. A strain sensitive tensile (Fig.1(b)) and linear shear softening model (taking into account dowel action and aggregate interlock) [7] is employed to simulate the post-cracking behaviour.

Reinforcing bars (or eventually prestressing cables) are considered as discrete steel bars or membranes of equivalent thickness [7]. The steel behaviour is therefore idealized using a uniaxial elasto-viscoplastic model resisting only the axial force in the bar (or the cable) direction.

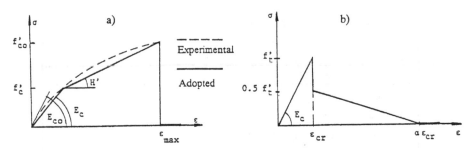

Figure 1. One-dimensional representation of (a) the compressive model and
(b) the tensile softening model for concrete.

The material modelling is primarily developed for ultimate load analysis. Nevertheless the model has proved to be reliable and efficient to the solution of structural dynamic problems [8-10].

Finite element formulation and solution procedure

The BET50 package is based on the 8-node rectangular serendipity isoparametric elements which enable the solution of 2D and axisymmetric problems. The "reduced" integration rule (i.e. 2x2) is exclusively employed. The rectangular elements represent the concrete structures while special bar or membrane elements, embedded into the rectangular ones are employed to simulate reinforcing bars or prestressing cables (tendons) in the x-y plane or in the (r-z) meridional and circumferential directions. The progress of cracked zones, as well as the nonlinear compressive behaviour of concrete is analysed and monitored for each element integration point. Geometric nonlinearities can be also taken into the account; a total Lagrangian approach is adopted. An incremental and iterative modified Newton-Raphson algorithm is adopted for the analysis of concrete structures under both monotonic static and transient thermal loading conditions. For the dynamic analysis the Newmark's implicit-explicit algorithm in a predictor-corrector form is adopted [14].

NUMERICAL EXAMPLES

Three numerical examples are presented which illustrate both the applicability of the BET50 package for the static, cyclic and dynamic analysis and also provide a comparison between numerical and experimental results.

Example 1: R.C. shear wall

The R.C. shear wall considered in this example correspond to one from the series of small scale walls tested experimentally at Tsinghua University, Beijing [15]. Details of the wall, together with the reinforcement arrangement, are given in Fig.2. Young's modulus, uniaxial compressive and tensile strength, Poisson ratio, and ultimate compressive strain were obtained from material testing [15] so that the one-dimensional representation of the constitutive relationships in compression (Dashed line in Fig.1(a)) and tension (Fig.1(b)) are accordingly established. The corresponding numerical model is then determined and parameters adopted are listed in Table 1. The tension and shear softening parameters are chosen from the spectrum of recommended values for the model employed [5].

Table 1: Problem parameters for Examples 1 and 2.

	Example 1	Example 2
CONCRETE		
Young's modulus, E_c	$3.1\ 10^7\ kN/m^2$	$2.6\ 10^7\ kN/m^2$
Poisson's ratio, ν	0.25	0.25
Compressive strength, f_c'	$30200\ kN/m^2$	$15400\ kN/m^2$
Tensile strength, f_t'	$2840\ kN/m^2$	$2103\ kN/m^2$
Tension softening coef., α	5	5
Shear softening coef., γ	10	10
Ultimate comp. strain, ϵ_{cu}	0.003	0.0024
REINFORCEMENT		
Young's modulus, E_s	$2.18\ 10^8\ kN/m^2$	$2.18\ 10^8\ kN/m^2$
Yield stress, σ_y	$2.88\ 10^5\ kN/m^2$	$3.12\ 10^5\ kN/m^2$
Hardening Parameter, H'	$1.91\ 10^6\ kN/m^2$	$3.11\ 10^5\ kN/m^2$

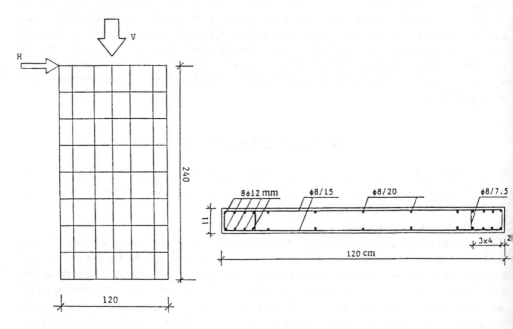

Figure 2. R.C. concrete wall (Example 1) showing finite element idealization and the wall cross-section.

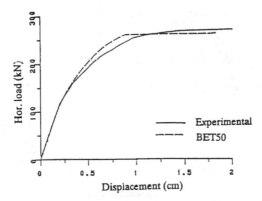

Figure 3. Comparison of numerical and experimental horizontal displacements with increasing horizontal loading for Example 1.

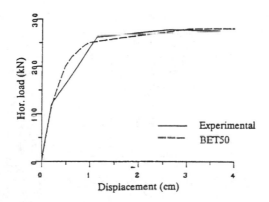

Figure 4. Numerical and experimental envelopes due to cyclic horizontal loading for Example 1.

The wall is first analysed under the conditions of a constant vertical load (V=400 kN) and monotonic increasing horizontal force in order to ascertain the parameters adopted. In Fig.3 the experimental load-horizontal displacement curve at the top of the wall is compared with the results obtained using BET50. Excellent agreement between the two results is evident.

The wall was then reanalysed using cyclic loading (horizontal force-controlled cycles). Typical force-displacement hysteresis loops for the top of the wall are shown in Fig.5. Numerical response (Fig.5(b)) is rather inaccurate, paricularly in the ultimate range, compared with the experimental response (Fig.5(a)). However, the numerical and experimental cyclic envelopes are seen to be in good agreement as shown in Fig.4. The deformed wall profiles and the development of zones of tensile cracking (dashed lines represent closed cracks at the particular moment) and compressive crushing are illustrated in Fig.6.

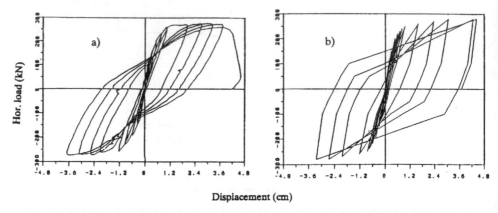

Figure 5. Cyclic response (Example 1): (a) experimental, and (b) numerical.

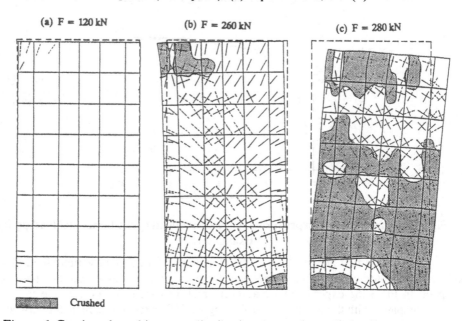

Figure 6. Crack and crushing zone distribution due to the cyclic loading at some significant phases (Example 1): (a) first cracks appear, (b) beginning of reinforcement yielding, and (c) failure.

Example 2: R.C. shear wall with openings

The R.C. wall with openings whose details are shown in Fig.7 is analysed in the same manner as Example 1. In this example the constant vertical load is V=83 kN. Concrete constitutive relationships are established according to the Chinese code of practice since only the concrete cube strength was known [15]. Therefore the numerical model and corresponding parameters listed in Table 1 have rather approximate values. Fig.8 compares the horizontal displacements of the top of the wall with the experimental and numerical results due to monotonic loading. Cyclic response, i.e. numerical and experimental hysteresis envelopes are given in Fig.9. The numerical response obtained is generally acceptable. Deviations (for higher loads approx. 10%) from the experimental results [15] can be attributed to the lack of complete material data and to a certain degree of unreliability of the parameters adopted.

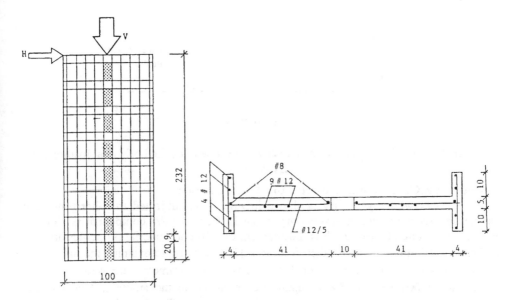

Figure 7. R.C. concrete wall with openings (Example 2) showing finite element idealization and the wall cross-section.

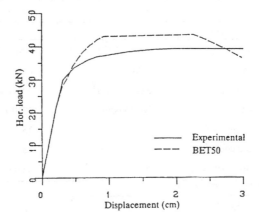

Figure 8. Comparison of numerical and experimental horizontal displacements with increasing horizontal loading for Example 2.

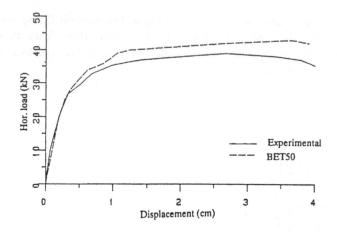

Figure 9. Numerical and experimental envelopes due to cyclic horizontal loading for Example 2.

Example 3: R.C. shear wall subjected to a base excitation

The R.C. wall whose geometry is shown in Fig. 10 is subjected to a horizontal base excitation (Accelerogram corresponding to the El Centro N-W component is shown in Fig.11). The finite element mesh is illustrated in Fig. 10 and material properties are listed in Table 2. All applied data is taken from Ref. [16]. In order to examine the influence of tension and shear softening parmeters [7] two sets of the parameters are used. A time step length $\Delta t = 0.001$ sec and 5% damping are employed in the analysis. The horizontal displacement response of the three wall levels is shown in Fig. 12, where the results are compared with results from [16].

Table 2: Problem parameters for Example 3.

	Set A	Set B
CONCRETE		
Young's modulus, E_c	$2.618 * 10^7$ kN/m^2	$2.618 * 10^7$ kN/m^2
Poisson's ratio, ν	0.20	0.20
Compressive strength, f_c'	32520 kN/m^2	32520 kN/m^2
Tensile strength, f_t'	2810 kN/m^2	2810 kN/m^2
Tension softening coef., α	20	15
Shear softening coef., γ	10	25
Ultimate comp.strain, ϵ_{cu}	0.003	0.003
REINFORCEMENT		
Young's modulus, E_s	$1.998 * 10^8$ kN/m^2	$1.998 * 10^8$ kN/m^2
Yield stress, σ_y	$3.686 * 10^5$ kN/m^2	$3.686 * 10^5$ kN/m^2
Hardening param., H'	0.0	0.0

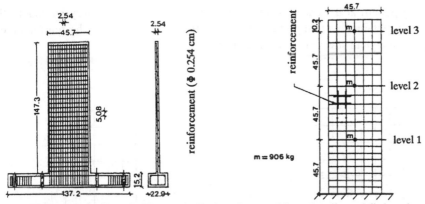

Figure 10. R.C. wall (Example 3) showing problem details and finite element idealization.

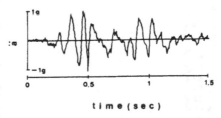

time (sec)

Figure 11. Accelerogram for Example 3.

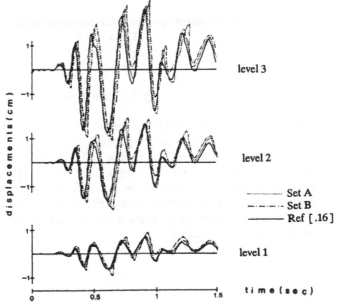

Figure 12. Horizontal displacement response of Example 3 for three levels.

The cracking pattern at the end of excitation shown in Fig. 13(a) is very similar to the experimentally obtained pattern presented in Ref.[16]. Positions where the reinforcement stress exceeds the yield stress are indicated in Fig. 13(b).

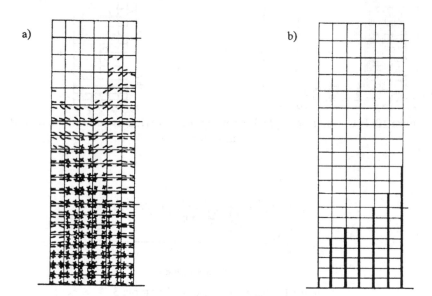

Figure 13. (a) Cracking pattern and (b) reinforcement yielding at the end of excitation
(Example 3)

CONCLUSIONS

The numerical examples presented illustrate the practical applicability of the numerical model and the BET50 finite element package developed, with good agreement with experimental results being generally evident. The nonlinear material model employed, although primarily developed for the ultimate load analysis, proved to be reliable as well as the solution process to be robust for the static, cyclic and dynamic analysis. It is shown that even in a case of a lack of complete material data an acceptable solution for practical engineering design can be reached. Knowing concrete cube strength as the only available data, valuable results are obtained.

REFERENCES

1.   Damjanic, F.B. et al.(Eds.), <u>Computer-Aided Analysis and Design of Concrete Structures</u>, Pineridge Press, Swansea, 1984.

2. Bicanic, N., Mang, H. (Eds.), Computer-Aided Analysis and Design of Concrete Structures, Pineridge Press, Swansea, 1990.

3. Owen, D.R.J., Damjanic, F., A finite element formulation for transient thermal analysis of nonlinear material problems, Trans. 7th SMiRT, Chicago, 1983, H, pp. 169-176.

4. Owen, D.R.J., Figueiras, J.A., Damjanic, F.B., Finite element analysis of reinforced and presstresed concrete structures including thermal loading. Comp. Meth. Appl. Mech. Engng., 1983, 41, 323-366.

5. Damjanic, F.B., Owen, D.R.J., Practical considerations for modelling of post-cracking concrete behaviour for finite element analysis of reinforced concrete structures. In Computer-Aided Analysis and Design of Concrete Structures, eds. F.Damjanic et al., Pineridge Press, Swansea, 1984, pp. 693-706.

6. Damjanic, F.B., A finite element formulation for analysis of reinforced concrete structures. In Finite Elements in Computational Mechanics - FEICOM/85, ed. T.Kant, Pergamon Press, 1985, pp. 413-422.

7. Damjanic, F.B., A finite element technique for analysis of reinforced and prestressed concrete structures. In Finite Element Methods for Nonlinear Problems, eds. P.Bergan, K.J.Bathe, W.Wunderlich, Springer, Berlin, 1986, pp. 623-637.

8. Damjanic, F.B., Mihanovic, A., Jaramaz, B., A two-dimensional finite element model for seismic analysis of reinforced concrete structures. In Computational Plasticity, eds. D.R.J.Owen et al., Pineridge Press, Swansea, 1987, pp. 1375-87.

9. Damjanic, F.B., Radnic, J., Seismic analysis of fluid-structure interaction including cavitation. In Computer Modelling in Ocean Engineering, eds. B.A.Schrefler, O.C.Ziekiewicz, Balkema, Rotterdam, 1988, pp. 523-530.

10. Radnic, J., Damjanic, F.B., 2D model for dynamic analysis of reinforced concrete structures. In Proc. 4th Symp. Comp. Civil Engng., ed. J.Duhovnik, University of Ljubljana, FAGG-IKPIR, Ljubljana, 1988, pp, 148-153 (In Croatian with English summary).

11. Damjanic, F.B., Finite element analysis of buried reinforced concrete pipes under internal pressure. NUMETA/85, eds. J.Middleton, G.Pande, Balkema, Rotterdam, Netherlands, 1985, pp. 679-686.

12. Mihanovic, A, Damjanic, F.B., Numerical analysis of nonlinear dynamic response of reinforced concrete structures. In Numerical Methods for Non-linear Problems, eds. C.Taylor et al., Pineridge Press, 1986, 3, pp. 1238-47.

13. Stanek, M, Damjanic, F.B., Celcer, V., Cracking prediction of a prestressed concrete septic containment. In Computer-Aided Analysis and Design of Concrete Structures, eds. N.Bicanic, H.Mang, Pineridge Press, UK, 1990, pp. 623-631.

14. Hughes, T.J.R., Pister, K.S., Taylor, R.L., Implicit-explicit finite elements in nonlinear transient analysis. Comp. Meth. Appl. Mech. Engng., 1979, **17/18**, 159-182.

15. Y. Zhou, Test investigation of the effect of section shape and reinforcement on seismic behaviour of the isolated walls. Research Institute of Structural Engineering, Tsinghua University, Beijing, China, 1988.

16. Agrawal, A.B., Jaeger, L.G, Mufti, A.A, Response of RC shear wall under ground motions, ASCE, J. Struct. Div., 1981, 395-411.

NONLINEAR SEISMIC ANALYSIS OF HYBRID REINFORCED CONCRETE FRAME WALL BUILDINGS

Hugo Bachmann, Thomas Wenk, and Peter Linde
Swiss Federal Institute of Technology (ETH), Institute of Structural Engineering
CH-8093 Zurich, Switzerland

ABSTRACT

A nonlinear dynamic analysis method to predict the ductility demand of reinforced concrete frame wall buildings under earthquake action is presented. The method is based on a general purpose finite element code enhanced with newly developed user defined elements. The plastic hinge zones in the wall and frame members are modeled by these user elements.

As numerical examples, four different capacity designed six-storey reinforced concrete buildings are subjected to a nonlinear time history analysis to determine the rotational member ductility demand.

INTRODUCTION

The capacity design method provides a clear and simple way for obtaining the earthquake resistance of reinforced concrete structures. The key steps of the method are the predetermination of the plastic hinge zones in the structure and the proper detailing of these, as well as the protection of the elastic parts of the structure against yielding [1]. For the numerical analysis of reinforced concrete structures under seismic action a software tool is necessary which properly simulates the hysteretic behaviour of the plastic hinges in walls and frames. A thorough evaluation of existing programs led to the development of macro hinge elements and their incorporation in a finite element package [2]. The purpose of this paper is to show the applicability of the new tool by seismic analyses of a series of two-dimensional frame wall buildings.

MODELLING OF STRUCTURAL WALLS

General Concepts
The numerical modelling of structural walls in this paper is based upon a so called macro model which attempts to simulate the global behaviour over a certain height of a wall. The original macro model was introduced in [3], and further modifications were suggested among others

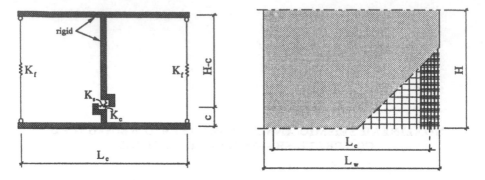

Figure 1. Macro model (left), simulating behaviour over height H of structural wall (right).

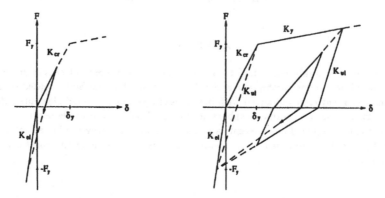

Figure 2. Hysteretic rules for flexural springs K_f in fig. 1 of wall model. Small excursions (left), and large excursions (right).

in [4-7]. The macro model used in this paper is shown in fig. 1. It consists of four nonlinear springs connected by rigid beams. The two outer springs K_f model the flexural behaviour, and the horizontal spring K_s models the shear behaviour. The vertical central spring K_c, finally, models the elastic axial behaviour together with the two outer springs, in compression only. Recently, some attempts were made to combine the four springs shown in fig. 1 (left) into a single two dimensional macro wall element. This element can be used for the modelling of intended wall plastic hinge regions, see e.g. [1], as well as for other regions of a wall. The possibility exists to discretize regions of a wall into several elements, as pointed out e.g. in [5]. This is considered as well in the modelling for this paper, although the bending moment variation with height for the tested walls (see section numerical examples), proved to be relatively uniform for the three lowest stories, due to participation of higher modes. Since no significant moment gradient occured over the element height, the distance c, shown if fig. 1, was kept to zero as in [3]. The walls studied in this paper have rectangular cross sections, and flexural reinforcement concentrated in confined zones at the ends. The location for the flexural springs of the model is taken at each centroid of flexural reinforcements, giving a central distance between the flexural springs of L_c, somewhat shorter than the wall length L_w, as seen in fig. 1 (right).

Hysteretic Models

The three types of springs in the numerical model each employ their own hysteretic model. The flexural springs together simulate the flexural behaviour of the entire wall section, i.e. also including the contribution of the web between the confined end zones. In this manner it is possible to achieve a reasonable simulation of the flexural behaviour as well as of the pure axial behaviour, in connection with the central vertical spring.

The hysteretic model for the flexural springs is shown in fig. 2, and is somewhat simpler than the one used in [3]. A moment-curvature analysis of the wall cross section is used for the determination of the the yield moment, as discussed in [5]. The yield moment is then transformed into a yield force for the flexural spring. In between yielding and compressive elastic behaviour, cracked stiffness is assumed as 70 % of elastic, at this time regarded as reasonable due to the poor ability of a moment-curvature relation to inform about crack and strain distribution over the element height. The tensile elastic stiffness of the first cycle is disregarded. The yielding stiffness is chosen as 3 % of elastic to account for some strain hardening. Together, the elastic, cracked and yield stiffnesses provide the skeleton curve. Before yielding, unloading may occur to a point on the elastic compressive branch corresponding to yield force level, as shown in fig. 2 (left). After yielding, unloading always at first occurs with a stiffness K_{ul}, shown in fig. 2 (right). At zero force level, the unloading takes a direction towards the yield force level on the elastic compressive branch. Reloading occurs towards the maximum reached point on the yielding branch, as shown in fig. 2 (right).

The here studied walls behaved mainly elastically in shear, although some minor shear cracking occured at some instance. Since shear played no major role, a bilinear origin oriented hysteretic model, as used in [3,5] however without yielding branch is considered sufficient. The elastic shear stiffness is obtained from theory, and cracking force as well as a cracked stiffness of 16 % of elastic, obtained according to empirical relations used in [6]. The central vertical spring, finally, employs a bilinear stiffness model, with compressive elastic stiffness obtained to achieve elastic behaviour under gravity load, and in tension with a stiffness close to zero.

MODELLING OF BEAMS

In reinforced concrete frames designed according to the capacity design method, plastic hinges will form only at predetermined and specially detailed locations. In the analysis these plastic hinge zones are modeled by a newly developed two node beam element with nonlinear hysteretic behaviour (fig. 3). The element length L_p is taken equal to the ductile detailed length of the beam in the structure. The remainder of the beam is modeled by linear beam elements.

A linear moment gradient is assumed over the length of the plastic hinge element. The stiffness EI of the element is a function of the average moment M_i (fig. 3) and is kept constant over the element length. A simplified Takeda hysteretic model with an asymmetric bilinear skeleton curve is used for the moment vs. curvature relation as shown in fig. 4 (left) [8]. It is important to include the effect of yield moment asymmetry in the beam hinge model of gravity load dominated frames.

The advantage of the simplified Takeda model is that the moment curvature relation is completely defined by three design parameters of the cross section: positive yield moment M_y^+, negative yield moment M_y^-, and elastic stiffness, as well as yielding stiffness taken as 3 % of elastic. The often observed phenomena of cyclic behaviour of reinforced concrete sections such as strength degradation, pinching and bond slip of reinforcement are undesirable in a capacity method designed structure and should be avoided as much as possible by appropriate constructive measures [1]. Consequently these phenomena are not included in the hysteretic

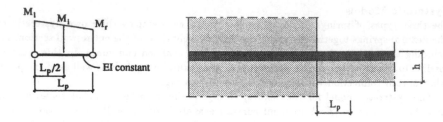

Figure 3. Macro model (left), simulating behaviour over length L_p of beam hinge (right).

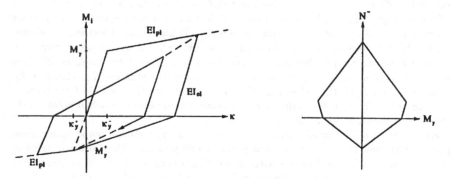

Figure 4. Hysteretic rules for beam plastic hinge (left), yield moment-axial force relation for column plastic hinge (right).

model. The influence of concrete cracking is taken into account from the beginning by reducing the elastic stiffness of both the hinge and the linear elements to 40% of the stiffness of the uncracked section.

MODELLING OF COLUMNS

Plastic hinges in columns are avoided in general (strong column – weak beam concept). However, at the foundation level the formation of plastic hinges in the columns can usually not be prevented. In addition plastic column hinges are often provided at the top floor, where normal forces in the columns are small.

The modelling of the column hinges is similar to the modelling of the beam hinges. Over the plastic hinge length L_p the column is discretized by a nonlinear hinge element as in fig. 3 (left), for the rest of the column a linear beam element is used. To account for the influence of the axial force, the skeleton curve of the column hinge element is expanded or shrunk as function of N. The yield moment-axial force relation used is shown in fig. 4 (right). The axial stiffness of the column hinge element is always kept elastic. Concrete cracking is taken into account by reducing the elastic stiffness of both the hinge and the linear elements to 80% of the stiffness of the uncracked section.

NUMERICAL EXAMPLE

Description of six-storey building designs

As numerical examples four different two-dimensional structures as shown in figs. 5 and 6 were used. Designs W and F are designed according to the Swiss codes SIA 160 and SIA 162 and the capacity design method [1]. Designs FW1 and FW2 are combinations of the structural elements of design F and W for the purpose of numerical comparisons.

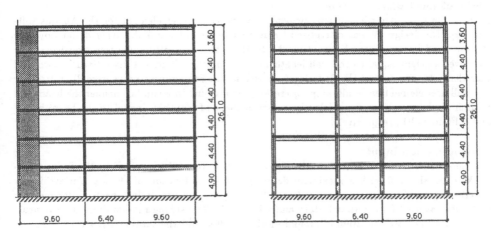

Fig. 5. Elevation of design W (left) and design F (right)

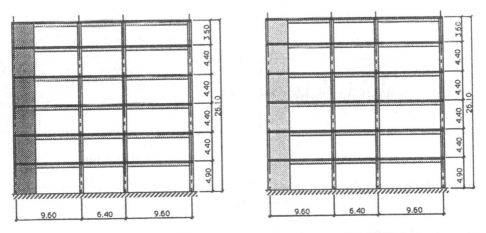

Fig. 6. Elevation of design FW1 (left) and design FW2 (right)

Design W: consists of a structural wall combined with gravity load columns designed for gravity load tributary to one bay as in design F, however for masses tributary to two transverse bay widths of 6.40 m. All the gravity load columns and beams are neglected in the time history analysis.

Design F: consists of a moment resisting frame designed for gravity load and masses tributary to one transverse bay width of 6.40 m. Plastic hinges are allowed to form in the beams at column faces, and in the columns at the foundation and the roof only.

Design FW1: is a combination of design W and design F. The structural wall of design W is combined in its plane with a moment resisting frame as in design F. The same gravity load and masses as in design W are assumed.

Design FW2: is equal to design FW1 except that the wall itself is designed for masses tributary to one bay only. In the time history analysis of design FW2, masses tributary to two bays are considered as in design FW1 and W.

Finite element discretization

Each of the three described macro elements was coded as user element for the Abaqus software [9]. The design W was discretized by user elements only, and the other designs by user elements and library linear beam elements. In the walls, a plastic hinge was modeled at the base over a height equal to the wall length L_w. The plastic hinge was discretized by two wall user elements. The rest of the first storey, as well as the remaining storeys were discretized by one wall user element each, allowing for cracking behaviour. In a similar manner the beam and column plastic hinges were modeled by a user element of length L_p equal to the beam height and column width, respectively.

Ground motion input

The analyzed building is located in the highest seismic zone (3b) specified by the Swiss earthquake code SIA 160 with a maximum design ground acceleration of 16 % g. An artificially generated ground motion compatible to the SIA 160 code elastic design spectrum for medium stiff soils was used for the time history analysis (fig. 7). The strong motion duration is approximately 7 s, the total duration of ground motion is 10 s, and the total analysis time is 12 s in increments of 0.01 s. The ground motion was applied horizontally in the plane of the frames. The dynamic analyses were preceded by an elastic static gravity load step.

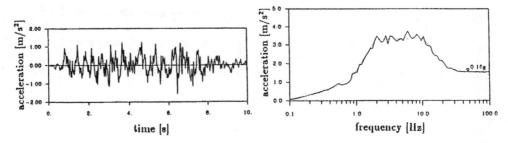

Figure 7. Time history of ground motion (left), response spectrum of ground acceleration for 5% damping (right)

Discussion of results

In figs. 8 and 9 the distribution of plastic deformations is shown by circles indicating the maximum rotational ductility demand during the 12 s time history analysis. For the wall plastic hinge the rotational ductility is shown individually for the upper and lower region and for left and right tension yielding. A maximum rotational ductility of 3.2 (right tension yielding) is obtained in the wall hinge of design W. Due to frame-wall interaction this value dropped below yielding in design WF1 (no circles shown) and attained a value of 2.2 (left tension yielding) in design WF2. In the frame hinges, the hybrid designs showed a higher ductility demand at the face of the wall (maximum rotational ductility of 7.8) compared to the design F, wheras in the rest of the frame it was lower.

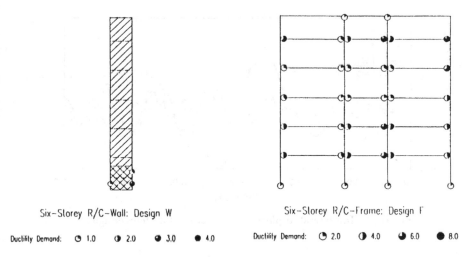

Figure 8. Distribution of rotational ductility demand of R/C-wall of design W (left) and R/C-frame of design F (right)

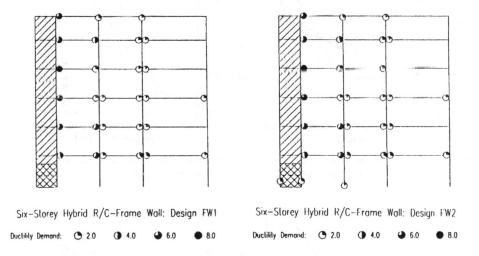

Figure 9. Distribution of rotational ductility demand of R/C-frame wall of design FW1 (left) and design FW2 (right)

Top lateral displacement histories are plotted in fig. 10. The displacement histories of the unsymmetric designs FW1 and FW2 were zeroed after the static gravity preload. Design W reached a maximum displacement of 210 mm (0.8 % of height), design F a maximum of 134 mm (0.5 % of height) and both hybrid designs FW1 and FW2 a maximum of 100 mm (0.4 % of height). The lateral displacement of the hybrid designs coincide up to a time of about 9 s.

For the wall the moment-curvature behaviour is shown in fig. 11 (left) for design W and in fig. 12 for the hybrid designs. The values are taken from the hinge element closest to the base. The maximum value of fig. 11 (left) corresponds to a rotational ductility of 3.2 as mentioned

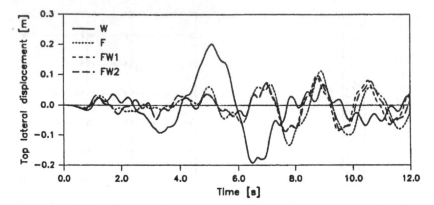

Figure 10. Top lateral displacement histories of all four designs

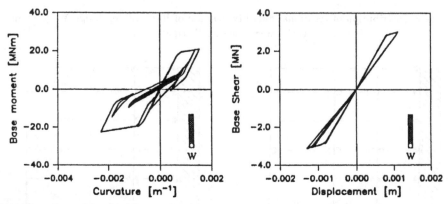

Figure 11. Moment-curvature behaviour (left) and shear-lateral displacement behaviour (right) of plastic hinge at wall base of design W

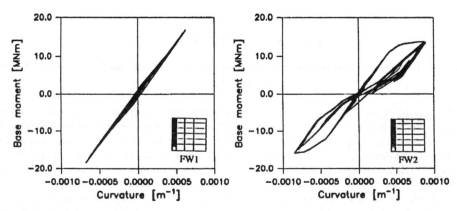

Figure 12. Moment-curvature behaviour of plastic hinge at wall base of design FW1 (left) and design FW2 (right)

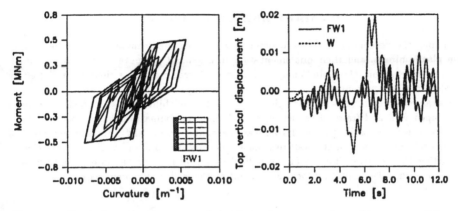

Figure 13. Typical moment-curvature behaviour of top storey beam hinges at wall face of design FW1 (left) and vertical displacement histories of inner wall edge at the same location for designs FW1 and W (right)

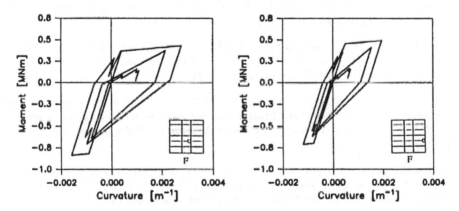

Figure 14. Typical moment-curvature behaviour of second floor beam hinges at interior column (left) and exterior column (right) of design F

above. Design FW1 did not reach yielding, the small nonlinearity shown in fig. 12 (left) being due to flexural cracking.

The shear behaviour (spring K_s of fig. 1) is shown in fig. 12 (right) for the design W showing some shear cracking, while the designs WF1 and WF2 behaved uncracked in shear. As an example of the high ductility demand at the wall face the moment-curvature behaviour of the roof beam hinge is shown in fig. 13 (left) for design FW1. The reduction of the wall uplift at the inner edge due to the frame wall interaction is shown for the same location in fig. 13 (right). Typical moment-curvature behaviour for the second storey beams of design F are shown in fig. 14. The rotational ductility demand of these beam hinges are 6 at the interior and 4 at the exterior beam.

SUMMARY AND CONCLUSION

In this paper, the development of macro elements modelling the behaviour of wall, beam, and column plastic hinges, and their implementation in a general computer code was described. As a numerical example, the global analysis of four reinforced concrete structures, modeled by macro elements, was presented.

The examples presented, served as a first check on the reliability of the interaction between the developed wall and frame user elements. Although no comparison basis, such as experimental data, was available, the results appear reasonable. Especially, the reducing effect of the frame-wall interaction on the wall rotational ductility demand and wall uplift are clearly seen. Further the positive influence of the hybrid design in limiting the lateral displacement and beam ductility demand at all locations except at the wall face, is demonstrated.

REFERENCES

1. Paulay T., Bachmann H., Moser K., Erdbebenbemessung von Stahlbetonhochbauten, Birkhäuser Verlag, Basel-Boston, 1990.

2. Wenk T., Bachmann H., Ductility demand of 3-D reinforced concrete frames under seismic excitation. Proceedings of the european conference on structural dynamics Eurodyn'90, A.A. Balkema, Rotterdam, 1991, Vol. 1, pp. 537-41.

3. Kabayesawa T.H., Shiohara S., Otani S., Aoyama H., Analysis of the full-scale seven-story reinforced concrete test structure. Proceedings, 3rd joint technical coordinating committee, U.S.-Japan cooperative earthquake research program, Building Research Institute, Tsukuba, 1982.

4. Vulcano A., Bertero V.V., Analytical models for predicting the lateral response of R C shear walls, evaluation of their reliability. Report No. UCB/EERC-87/19, Earthquake Engineering Research Center, University of California, Berkeley, 1987.

5. Linde P., Experimental and analytical modeling of R/C shear walls. Engineer's Thesis, Stanford University, Stanford CA, 1988.

6. Linde P., Analytical modeling methods for R/C shear walls. Report TVBK-1005, Lund Institute of Technology, 1989.

7. Fajfar P., Fischinger M., Mathematical modeling of reinforced concrete structural walls for nonlinear seismic analysis. Proceedings of the european conference on structural dynamics Eurodyn'90, A.A. Balkema, Rotterdam, 1991, Vol. 1, pp. 471-478.

8. Takeda T., Sozen M.A., Nielsen N.N., Reinforced concrete response to simulated earthquakes. Journal of the Structural Division, ASCE, New York, 1970, Vol. 96, pp. 2557-73.

9. "ABAQUS User's Manual". Hibbitt, Karlsson & Sorensen, Providence, Rhode Island, 1990.

ANALYSIS OF ULTIMATE STATES OF REINFORCED CONCRETE WALL - FRAME STRUCTURES

JURE BANOVEC
University of Ljubljana, Department of Civil Engineering,
Jamova 2, Ljubljana, Slovenia

ABSTRACT

The objective of this paper is to present a finite element formulation of planar reinforced concrete (RC) frames with structural walls. A mathematical model of a structural (shear) reinforced concrete (RC) wall is introduced. It consists of the already verified "axial-flexural" element P_4 (for Bernoulli's beam), in which the interpolation functions for flexural and axial deformations are based on a fourth order polynomial and of nonlinear "shear" springs that model include the shear deformations. The numerical examples show the applicability of the proposed model.

INTRODUCTION

Building structures are usually modelled using beam-column elements. Many of the known methods divide the frame beams and columns into a large number of elements in order to achieve accurate predictions. As a result, a mathematical model with a very large number of degrees of freedom is to be analysed. The author has developed an effective method for the nonlinear static analysis of planar frames, which decreases the number of elements and provides results of adequate accuracy.

DESCRIPTION OF THE METHOD

The method is implemented in the computer program **NONFRAN**, which analyses the elastic plastic behaviour of plane frames by incremental Newton-Raphson method. The program can deal with the first order or the third order theory. In **NONFRAN** a Bernoulli's beam is used, which is characterised by axial and flexural deformations so that the cross section remains orthogonal to the beam axis. A beam is modelled with the P_4

element, which is described in /1/ and is only summarized briefly. The interpolation functions for flexural and axial deformations are based on the fourth order polynomial. Therefore, the results do not depend on the choice of the element axis and the element can be successfully applied for modeling the flexural and axial part of deformation of structural wall, where the neutral axis of the wall shifts towards the compression end when the horizontal loading increase.

During the incremental elastic plastic analysis (the spread of plastic zones), normal stresses across the cross-section are integrated. A typical RC cross-section is not homogeneous, because different stress-strain relationships, e.g. for longitudinal reinforcement, for concrete inside and concrete outside of transverse reinforcement, should be considered. Using the layer technique, the cross-section is divided in different rectangular subsections (subelements), which follow any nonlinear (piecewise linear) stress-strain relationship.

In order to realistically simulate the real behaviour of structural wall, the properties of the P_4 element have been improved by implementing the nonlinear shear springs (Fig. 1).

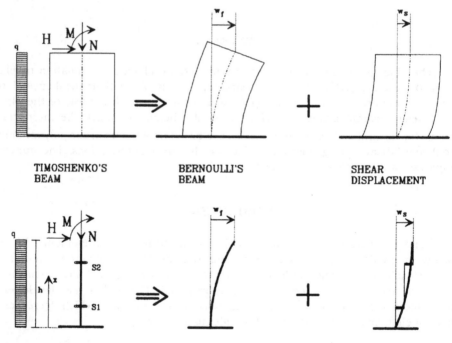

Figure 1. Mathematical model of the wall (Timoshenko's beam)

Let the shear force (Q) - shear deformation (γ) relationship be given and it can vary along the wall. We want to calibrate the shear springs. According to the principle of the complementary work we obtain

$$w_s = \int_0^h \gamma(Q)\, dx = h \int_0^1 \gamma\, d\xi \qquad (1)$$

$$\xi = \frac{x}{h}$$

Using numerical integration along the wall Eqn (1) may be written in the form

$$w_s = h \sum_{i=1}^{n} A_i \gamma(Q(\xi_i)) = \sum_{i=1}^{n} \Delta w_i \qquad (2)$$

n ... number of the integration points
A_i ... the weighting factor

Thus the Q_i - Δw_i relationship is obtained:

$$A_i\, h\, \gamma(Q(x_i)) = \Delta w_i \qquad (3)$$

If shear deformations do not vary along the wall, Eqn 2 becomes

$$h\, \gamma = \Delta w = w_s \qquad (4)$$

and the position of the shear spring does not influence the result.

SINGLE STORY ISOLATED WALL

We have studied numerically a simple RC wall with a rectangular cross section (Fig. 2).

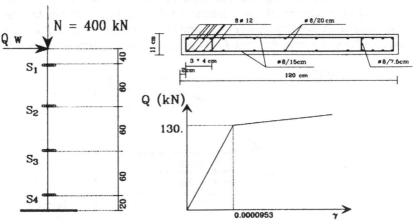

Figure 2. Mathematical model of RC wall

The RC wall was tested at Tsinghua University in Beijing /2/. The height of the wall is 2.4m, the cube strength of concrete is 36.0 MPa and the yield strength of reinforcement is either 381 MPa (ϕ 12 bars) or 288 MPa (ϕ 8 bars). The modulus of elasticity for concrete and steel are Ec = 31000 MPa and Es = 220000 MPa, respectively. The well known Kent-Park relationship was used for concrete in compression. Diagrams in Fig. 3 are used for concrete in tension and for steel reinforcement. The mathematical model adopted for the problem analysis had five beam column elements and four nonlinear shear springs. Nonlinear shear springs employed in the numerical model are controlled by a shear force - shear deformation diagram which was derived from the experiment.

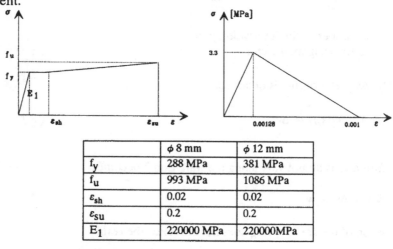

	ϕ 8 mm	ϕ 12 mm
f_y	288 MPa	381 MPa
f_u	993 MPa	1086 MPa
ε_{sh}	0.02	0.02
ε_{su}	0.2	0.2
E_1	220000 MPa	220000MPa

a) reinforcement   b) concrete

Figure 3. Idealized stress - strain relationships

First we have investigated the influence of the number of elements on the accuracy and convergence of the numerical solution. Fig. 4 shows the results of the adopted model with five elements (Fig. 2) and of the model with one P4 element and one shear spring. It can be seen that both Q - w relationships practically coincide.

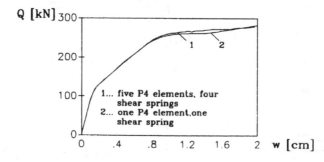

Figure 4. Shear force - horizontal top displacement relation for RC wall

Next, we present shear force - vertical top displacement curves (Fig. 5) for three different points. As observed, all three displacements rapidly increase in the range of the ultimate load. The great shifting of the neutral axis is present.

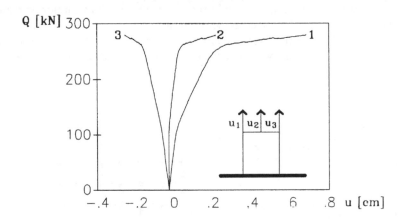

Figure 5. Shear force - vertical top displacement relationship

The numerical results are compared with the experiment in Fig. 6. Using experimental data for shear force - shear deformation relationship the numerical results show very good agreement with the experiment. Additionally we have investigated the influence of the shear spring hardening on the results (Fig. 6). Special attention should be paid to the determination of the appropriate percentage of hardening if experimental data are not available.

Finally, we tested the influence of the different choice of the element axis on the results. The comparison between the two axes (one in the middle of the wall cross section (curve 1b) and the other at the edge of the cross section (curve 1c)) is represented practically by the same curve.

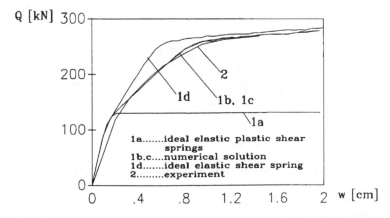

Figure 6. Shear force - horizontal top displacement relation ship

SEVEN STORY FRAME - WALL BUILDING

The seven story RC frame wall structure, which was tested in full-scale in Tsukuba, Japan, (Fig. 7) has been described in details in many papers related to the U.S. - Japan joint research project (e.g. in lit /3/). Only the most relevant information is summarized here. Typical column and beam cross sections are given in Fig. 8. The shear wall is reinforced with 2 ϕ 10 bars at spacing 20 cm in the horizontal and vertical directions. The concrete used in the structure was specified to have the 28-day cylinder strength of 27 MPa.

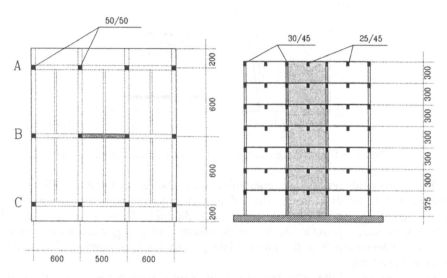

Figure 7. Plan and section of the building

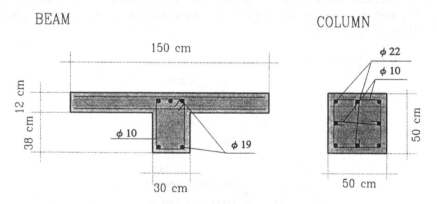

Figure 8. Cross-sections of a typical column and beam

The numerical model adopted for NONFRAN is shown in Fig. 9 . The structure was idealized as three parallel plane frames. Floor slab was assumed to be rigid in its own plane. It can be seen that the beam-columnn joints were simulated by rigid elastic elements. The total number of beam-column elements was 449 (the multilayer model) and number of shear springs was 14. The wall model had two shear springs over each story height. The structural system was analysed under a monotonously increasing lateral load with inverted triangular distribution. The characteristics of the shear force - shear deformation relationship for RC wall were used as in lit /3/. The used elastic shear rigidity is 1.02 E7 (kN) and the ultimate shear force is 3810 kN.

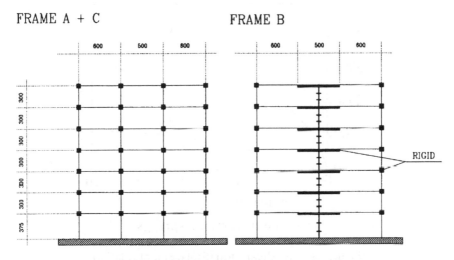

Figure 9. Mathematical model of 7 - story RC building

We have made many numerical calculations using different stress-strain relationships for concrete. The changing of the slope in the post ultimate range of stress-strain diagram did not (practically) influence the results. So, the presented results are obtained with zero slope in the post ultimate range. The outriggering provided to the wall-frame B by the frame A+C has been also numerically tested. The numerical model was adopted so that the adequate storey displacements of frames A+C and B were constrained by nonlinear vertical springs. Varying the characteristics of springs we have studied three-dimensional behaviour. The presented results have been obtained using the springs following the diagram in Fig. 10.

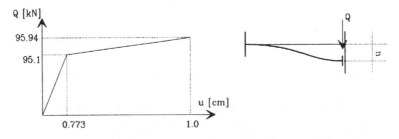

Figure 10. The characteristics of vertical springs

The results show (Fig. 11) that the outrigging action on the wall-frame B causes the displacement hardening of this frame. On the other side, this action causes the minimal displacement softening of frames A+C. Observing the whole system the three-dimensional behaviour causes the moderate displacement hardening.

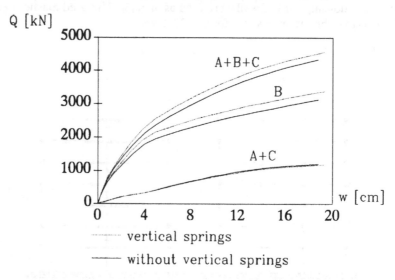

Figure 11. Base shear - top displacement relation

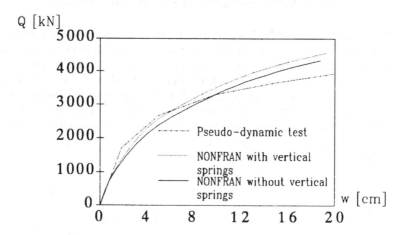

Figure 12. Comparison of top displacement and base shear relations

The numerical results are compared with the displacement envelope of the pseudo-dynamic test /3/. Considering the different approaches (monotonically increased loading versus the test PSD) a good correlation was achieved.

CONCLUSION

A mathematical model of a structural (shear) RC wall has been presented. It enables to take into account any nonlinear shear force - shear deformation relationship. The presented results show good agreement with the experiment, when the experimental data for the shear force - shear deformation relationship are available.

The presented model can be extended to include the influences of axial force and bending moment on the shear force - shear deformations relationship and will be addressed in future publications.

ACKNOWLEDGEMENT

This paper is based on work supported by the Ministry of Research in Slovenia and by the US. - Yugoslav Joint Fund for Scientific and Technological Cooperation.

REFERENCES

1. Banovec, J., An efficient finite element method for elastic - plastic analysis of plane frames. In Nonlinear Finite Element Analysis in structural Mechanics, Proceedings of the Europe - U. S. Workshop, Ruhr - Universitat Bochum, Germany, Springer Verlag, pp. 385 - 402, 1981.

2. Zhou, Y., Test investigation of the effect of section shape and reinforcement on seismic behaviour of the isolated wales. A thesis submitted for the degree of master of science. Research Institute of Structural Ingeneering, Tsinghua University, Beijing, 1988.

3. Kabeyasawa, T., Shiohara, H., Otani, S. and Aoyama, H., Analysis of the full - scale seven - story reinforced concrete test structure. Journal of the Fac. of Eng., Univ. of Tokyo, 1983, Vol 37, pp. 431 - 478.

LINEAR AND NONLINEAR SEISMIC RESPONSE OF PREFABRICATED COMBINED (FRAMES-TALL SHEAR WALLS) SYSTEM OF CONSTRUCTION

MEHMED ČAUŠEVIĆ
University of Banjaluka, Department
of Mechanical Engineering, Danka Mitrova 63A
78000 Banjaluka, Yugoslavia

ABSTRACT

On the base of the experimentally obtained results on full-scale structures and numerical analysis this paper gives some conclusions about the behavior of tall shear walls in the prefabricated combined system as well as the behavior of the system itself. It was concluded from the full-scale tests that only frames in the upper parts of this structure are participating in resisting forces induced by earthquakes. The mathematical modeling of these structures has also been performed. This included modeling of non-structural elements. It was created an additional vertical shear element simulating the influence of non-structural partition walls on the overall stiffness of the building. Analytical investigations included linear and nonlinear time-history analysis. A fair correlation of the results obtained analytically and experimentally was observed.

INTRODUCTION

In this paper the IMS prefabricated combined (frames-tall shear walls) structural system will be discussed. This system has been developed by the Institute of Material Testing of Serbia (IMS) from Belgrade and used for apartment building in the seismically active and nonactive regions in China, Italy, Angola, Egypt, USSR, Cuba, Ethiopia, Hungary and Yugoslavia. Because of its international use, IMS system has been chosen as one among topics in the three year research project sponsored by the U.S. - Yugoslav Joint Board /3/. This system consists of prefabricated reinforced concrete columns and floor diaphragms (slabs). The column-slab connection is attained by prestressing. It has been believed that horizontal loading is carried mainly by cast-in-place reinforced concrete shear walls of the types from Fig. 1 (even prefabricated shear walls are possible), which are either cantilever walls (Fig. 1a) or coupled by coupling beams representing floor slabs.

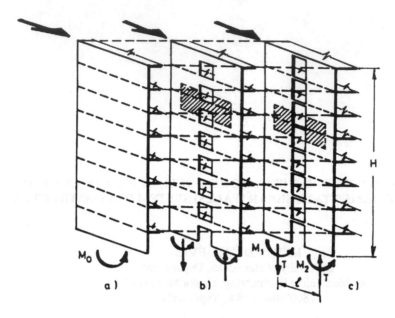

Figure 1. A schematic comparison of ductile tall shear walls:
(a) a cantilever wall, (b) walls coupled by strong beams,
(c) walls coupled by slabs only. (From Reference 2)

In this paper some further observations on all types of tall shear walls from Fig. 1 as the elements of the combined system of construction and on the combined system itself will be presented.

EXPERIMENTAL AND NUMERICAL LINEAR AND NONLINEAR ANALYSIS OF FULL-SCALE PREFABRICATED COMBINED SYSTEM OF CONSTRUCTION

Fundamental periods and modes of the structure presented in Fig. 2 have been obtained by using forced vibration testing of small amplitudes (Table 1). All tests have been performed by IZIIS Skopje. Viscous damping has been obtained by using half-power method. It has been ascertained about 1.7% of damping for this combined type of a structure in these tests.

TABLE 1
Fundamental periods and damping capacities obtained experimentally

Direction of excitation	T (sec)	ξ (%)
E-W	0.488	1.27 ... 1.71
N-S	0.407	0.93 ... 1.73
torsion	0.394	1.79 ... 3.42

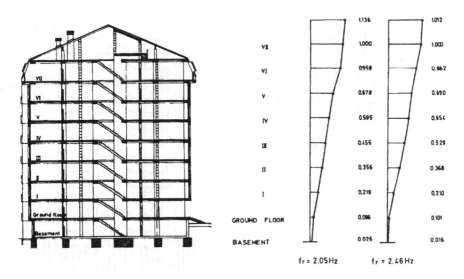

$f_r = 2.05\,Hz$ $f_r = 2.46\,Hz$

Figure 2. Vertical section and fundamental modes of the seven story apartment building built by prefabricated combined system of construction, E-W and N-S directions of excitation.

It can be concluded from the shapes of the fundamental modes, Fig. 2, that this structure has been made of uniform quality and good grade concrete, because there were no discontinuities in the measured shapes of the fundamental modes. From the shapes of these fundamental modes can be also concluded that the bending type of behavior of this structure will be prevailing only in the lower part of the structure. In the upper part of this structure the shear type of behavior is noticed. These experimental results proved previously theoretically obtained conclusion /7/ that forces induced by earthquakes have been carried by shear walls in the lower part and only by frames in the upper part of the building built by IMS system of construction. This is schematically shown in Fig. 3.

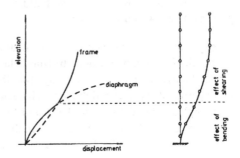

Figure 3. Schematic presentation of structure's behavior for prefabricated combined type of structure.

The second building (presented in Fig. 4) has been made by the same system of construction as the previous one and experimentally and numerically analysed.

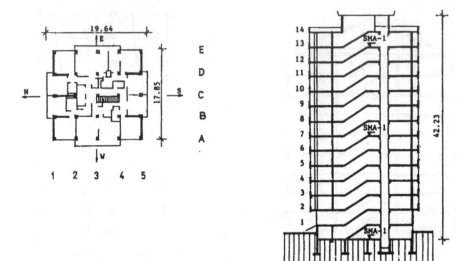

Figure 4. Floor-plan and vertical section in N-S direction of the 14-storey apartment building built by prefabricated combined system of construction (columns 38x38 cm, thickness of shear walls 15 cm), with the position of strong motion accelerometers.

This 14 storey apartment building had been chosen for the analysis because it possessed data which consists of:

- experimentally obtained dynamic characteristics by forced vibration testing on the building with very few non-structural elements, when masses are estimated to 59% of the final mass in the exploitation stage (the first test);
- experimentally obtained dynamic characteristics by forced vibration testing on the almost completed building with the majority of non-structural elements when masses are estimated to 81% of the final mass in the exploitation stage (the second test);
- instrumentation of this building with strong motion accelerometers, Fig. 4;
- registration of accelerations of the actual earthquake Banjaluka, August 13, 1981, M = 5.3, Fig. 8a;
- inspection of the building after this registered earthquake and detection of the damages when only minor non-structural damage was reported, mostly at the top of the building;
- experimental investigation of dynamic characteristics of the structure after this earthquake by ambient vibration tests in order to obtain dynamic changes of the structure bacause of its softening in this earthquake (the third test).

Some of the test results are presented in Table 2. All tests have been performed by IZIIS Skopje.

On the bases of all these data mathematical models were created. Firstly, a pseudo-tridimensional model was applied for the linear vibration analysis. Stiffness matrices have been computed for planar assemblages of shear walls and frames, Fig. 5.

TABLE 2
Experimentally and analytically obtained natural periods

		Pre-earth. Test 1		Pre-earth. Test 2		Banja Luka Earth.		Post-earth. Test	
		T1	T2	T1	T2	T1	T2	T1	T2
TEST	E-W	0.75	0.14-0.16	0.73-0.77	0.18	1.00-1.02	0.21-0.23	0.78	0.20
	N-S	0.79-0.80	0.20	0.72-0.77	0.20-0.21	1.10	0.23-0.24	0.82	0.22
	Tors.	0.70	0.16	0.64-0.67	0.16-0.17			0.71	0.17
ANALYSIS	E-W	0.79	0.14	0.75	0.16	0.94-1.03	0.18-0.19		
	N-S	0.86	0.18	0.76	0.19	1.07-1.12	0.24-0.26		
	Tors.	0.66	0.12	0.63	0.13				

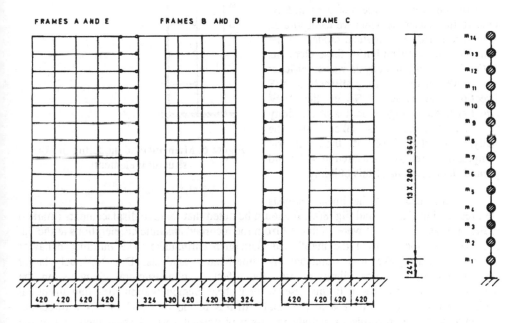

Figure 5. Mathematical model of the structure from Fig. 4 in N-S direction for stiffness, periods, mode shapes - linear analysis.

These frames have been formed by columns and floor slabs. After performing the static condensation procedure a dynamic model with 3x14 degrees of freedom was obtained. Because of the symmetry of the structure, Fig. 4, the translational and torsional vibrations are uncoupled and three separate planar analyses were performed, as shown in Fig. 5 for N-S direction. The computer program EAVEK /4/ has been used for all linear analyses. When calculating cross-section properties of shear walls the actual L shape of shear walls has been considered, Fig. 4. Also the equivalent beams (represented by floor slabs as shown in Fig. 1c) have been considered as effective.

Simulating the first forced vibration test performed on this 14-storey structure, only stiffness and mass of the bare structure have been considered in the free vibration analysis. Additional mass and stiffness were used to simulate this building with non-structural elements in the second forced vibration test. Experimentally determined fundamental periods of both translation vibrations were used to determine the cross-section properties of the additional vertical shear elements presented in Fig. 6 simulating the influence of non-structural partition walls on the overall stiffness of the building. This vertical shear elements are incorporated in the presented mathematical models of the building. The contribution of these elements in the free vibration analysis was significant (about 40% of the resulting stiffness in N-S direction and about 30% in the E-W direction). An average modulus of elasticity of $3.6 \times 10^5 \, MN/m^2$ was taken into the analysis on the base of the cylindric compressive strengths measured in walls and ground floor columns of this building.

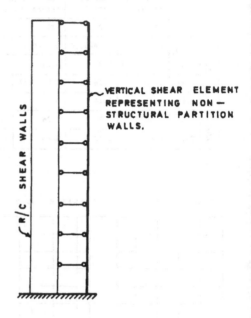

Figure 6. Mathematical modeling of non-structural elements

Some results of the linear analysis are presented in Table 2 and Fig. 8(a). It should be noted that relative displacements (marked in Fig. 8(a) as "observed") were computed from the recorded absolute accelerations at the 7th and the 13th floors during the mentioned Banjaluka earthquake. The building behavior during this earthquake was simulated by time-history analysis. Due to much larger deformations in the force vibration tests, the effect of non-structural elements on the structural stiffness was lower. In this case, vertical shear elements (simulating non-structural walls) contributed 8% and 18% to the final stiffness in the N-S and the E-W direction, respectively. It was possible to include in this way created vertical shear elements in the mathematical model using experimentally obtained results on the bare structure in the first forced vibration test. Time-history computations have been performed using different values of viscous damping. The best correlation with the observed response was obtained with 3.5% damping. This amount of damping is greater in comparison to the experimentally obtained values performed by using forced vibration small intensity testing (when 1% to 2% of critical damping have been obtained), which once again confirmed the statement about larger damping in the structure induced during an earthquake.

Within the framework of analysis cracking and yielding moments (M_{cr} and M_y) were established for cross-sections of all carrying elements of the structure. The comparison of these values with maximum bending moments obtained by the linear time-history analysis indicated that only local cracking occured in shear walls (in the first storey and from the 7th

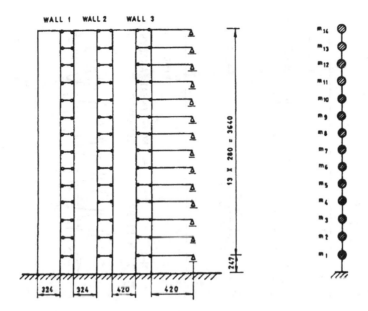

Figure 7. Mathematical model of the structure from Fig. 4 in N-S
direction for nonlinear analysis.

to the 10th storey) and in some columns from the 10th to the 14th storey. This local cracking
was more emphasized in the N-S direction. Therefore, a nonlinear dynamic analysis has been
performed in the N-S direction using the DRAIN-2D computer program /5/ and the
mathematical model presented in Fig. 7.

The results of these nonlinear analyses are presented in Fig. 8(b). A fair correlation of
the results obtained analytically and from the recorded absolute accelerations have been
observed.

CONCLUSIONS

When analysing and comparing experimental results obtained in the full-scale forced
vibration testing, results of integration of recorded absolute accelerations, results obtained in
the same structure by using ambient testing and mathematical modeling of the structure, the
following main conclusions have been obtained:

(1) The bending type of behavior typical for tall shear walls will be prevailing only in the
lower part of this structure made of prefabricated combined system of construction. In the
upper part of this structure the shear type of behavior is noticed. This means that in the upper
part of this structure only frames are carrying forces induced by earthquakes.

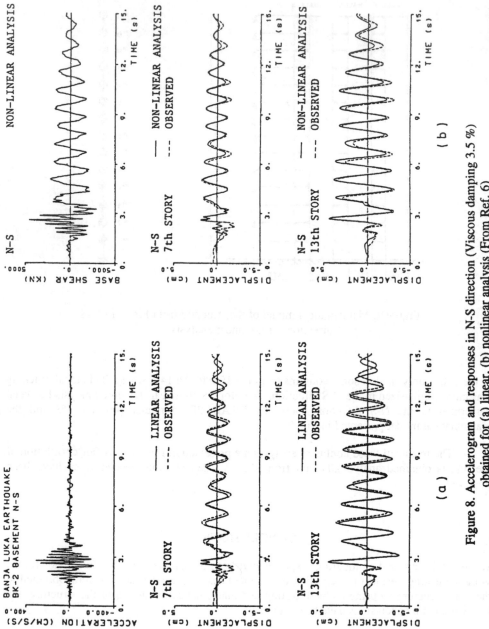

Figure 8. Accelerogram and responses in N-S direction (Viscous damping 3.5 %)
obtained for (a) linear, (b) nonlinear analysis (From Ref. 6)

(2) A good correlation of experimental and analytical results has been obtained in both linear and nonlinear analysis. However, this agreement would not have been attained if experimental results had not been used in mathematical modeling of non-structural elements. So far the additional vertical shear elements, simulating the influence of non-structural partition walls on the overall stiffness of the building, have been created.

ACKNOWLEDGEMENT

The results reported in this paper are based on work supported by the U.S.-Yugoslav Joint Fund for Scientific and Technological Cooperation, in cooperation with the NSF and the University of California, Berkeley, under grants NSF/JFP-519 and 525.

REFERENCES

1. Park R. and Paulay T., Reinforced Concrete Structures, John Wiley, 1975.

2. Paulay T., The Design of Reinforced Concrete Ductile Shear Walls for Earthquake Resistance, Research Report, University of Canterbury, Department of Civil Engineering, Christchurch, New Zealand, 1981.

3. Čaušević M., Mathematical Modeling of Reinforced Concrete Structures Subjected to Earthquakes, A Final Report to the U.S.-Yugoslav Joint Board on Scientific and Technological Cooperation, Project No. JFP 519, University of Banjaluka, 1988.

4. Fajfar P., EAVEK, A Program for Elastic Analysis of Multistorey Buildings, IKPIR Publication No. 13, University of Ljubljana, 1976 (in Slovene).

5. Kanaan A.E. and Powell G.H., DRAIN-2D, A General Purpose Computer Program for Dynamic Analysis of Planar Structures, Report No. UCB/EERC-73/6, University of California, Berkeley, 1973.

6. Fajfar P., Čaušević M. and Yi Jiang, Comparison of Analytically and Experimentally Determined Dynamic Behavior of a Multistory RC Building, Proceedings of the International Conference on Design, Construction and Repair of Building Structures in Earthquake Zones, The European Organization for the Promotion of New Techniques and Methods in Building EUROBUILD (Editor N. Hajdin), Dubrovnik, 1987.

7. Petrović B., International Scientific Symposium devoted to the use of IMS system of construction while building Olympic Village for the 14th Winter Olympic Games, (discussion), Sarajevo, 1986.

8. Čaušević M., Čorić B. and Srečković G., Experimental and Analytical Analysis of Prefabricated and Cast-in-Place RC Structures in Seismic Zones, Proc. of the 9th European Conf. on Earthq. Engrg., Moscow, 1990, Vol. 8, p.p. 275-284.

LABORATORY TESTING AND MATHEMATICAL MODELING OF RC COUPLED SHEAR WALL
SYSTEM FOR NONLINEAR STATIC AND DYNAMIC RESPONSE PREDICTION WITH
CONSIDERED TIME VARYING AXIAL FORCES

BORIS SIMEONOV
DANILO RISTIĆ
Institute of Earthquake Engineering and Engineering Seismology
University "Kiril and Metodij", Skopje, Yugoslavia

ABSTRACT

This paper presents some of the experimental results from recently
conducted laboratory tests up to failure of a large-scale six-storey RC
coupled shear wall specimen, and a newly proposed analytical model for
nonlinear static and dynamic response prediction of integral structures
with consideration of effectively induced time variation of axial forces.
The designed test specimen representing the lower part of the existing
high-rise building has been tested under simultaneous application of
reversed earthquake-like shear force and time-varying compressive axial
loads applied at top section centroid points of both left and right shear
walls. Typical, but very complex damage patterns of the specimen have been
recorded following simulated reversed and increasing shear displacement
amplitudes up to failure. To achieve realistic inelastic behaviour
simulation of the tested RC coupled shear wall system under complex and
arbitrary loading conditions, specific nonlinear multi-interface finite
element (NON-MIFE) model has been formulated and applied for prediction of
specimen's nonlinear static and nonlinear dynamic response.

INTRODUCTION

A six-storey model, representing the lower part of a high-rise building,
has been constructed for experimental testing. The studies are aimed at
determination of the hysteretic response of the model, evaluation of its
strength and deformability characteristics under cyclic loads, as well as
verification of the design criteria for construction of seismically
resistant RC shear walls. The first part of this paper presents the design
geometry of the model, the reinforcement distribution, and other
characteristics based on which the values of M-Φ relation were calculated
for various cross-sections of the model, to be used in analytical studies.
Finally, in the first part is included the hysteretic curve Q-Δ, recorded
during the experimental testing.

In the second, analytical part of this paper, presented and briefly discussed is the newly formulated nonlinear multi-interface finite element model (NON-MIFE) applicable for theoretical prediction of RC element nonlinear complex response under interactive effects of cyclic shear and time-varying axial forces, as well as implemented computational procedure for inelastic static and dynamic response analysis of the integral RC structures. To demonstrate the applicability of the proposed analytical model, an example nonlinear static and earthquake response analysis of the tested RC six-storey coupled shear wall specimen has been also included and discussed in this paper.

TEST SPECIMEN

The model dimensions and reinforcement distribution along the height are shown in Fig. 1. The constructed test model represents the lower part of a

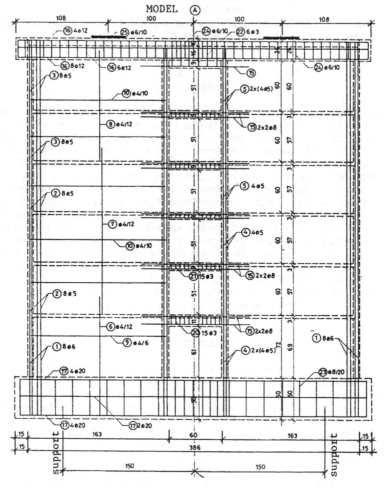

Figure 1. Geometrical properties and reinforcement distribution of the tested RC six-storey coupled shear wall specimen (CSWS)

high-rise building in scale 1:5. The height of the first storey is 72 cm, while the height of the remaining five storeys is 60 cm. In fact, the model consists of two RC shear walls strengthened with end column-wise vertical parts, integrated with coupling RC beams of 60 cm span.

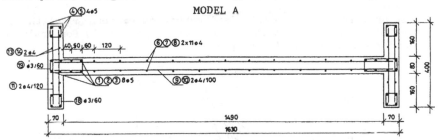

Figure 2. Cross-section and reinforcement distribution of specimen left and right shear walls integrated with coupling beams

The reinforcement distribution in the transverse cross-section of one of the walls is shown in Fig. 2. The reinforcement of wall both end vertical columns is concentrated at three points, with 4 or 8 longitudinal reinforcement bars , and transverse reinforcement (hoops) of ϕ 3mm/6cm. The percentage of the longitudinal reinforcement in both end vertical thicker parts, compared to their section area, is 1.36% for the first storey and 1.12% for the other storeys. The shear wall vertical reinforcement is assumed to be 0.26% for all the storeys, while the shear wall horizontal reinforcement is 0.52% for the first and 0.31% for the other storeys. The main reinforcement in the coupling beams consists of horizontal reinforcements, in the upper and lower part of the beam, and hoops of ϕ 3mm/4cm.

MECHANICAL PROPERTIES OF THE MODEL

The mechanical properties of the model are presented through the moment-curvature relation of walls cross-sections and coupling beams. In analytical determination of the bending moment and curvature, concrete modulus of elasticity was considered Ec = 30.5 GPa, concrete tension strength fct = 2.18 MPa, and steel yielding stress fsy = 400 MPa.

The M-ϕ relation has been calculated for three different response states representing occurrence of cracks due to bending, yielding of reinforcement under tension stress, and ultimate state (ultimate point – U). For the needs of analytical studies, the values of M-ϕ relation were calculated for all the elements cross-sections, but for the left and right shear wall, these values were calculated based on considered different levels of the axial compressive force in the walls.

When determining the moment and curvature for reinforcement yield state and ultimate state, reinforcement hardening was considered, and the concrete ultimate compressive strain was accepted to be εcu = 0.0035.

LABORATORY TESTING AND TEST RESULTS

The experimental testing of the model was performed with quasi-static cyclic loads applied at the top of the model. The gravity loads of the floors above, in model testing were simulated with two vertical forces applied at the top of the model. These axial forces were considered variable during testing depending on the horizontal cyclic load value.

The shear force - horizontal displacement curve for the sixth floor (Q-Δ) is shown in Fig. 3 and this envelope curve is a basis for analytical studies, and verification of the proposed model.

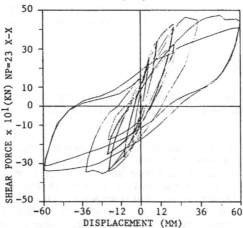

Figure 3. Experimentally recorded shear force - displacement hysteretic curve of the tested specimen under varying axial loads

FORMULATION AND VERIFICATION OF THE INELASTIC NON-MIFE MODEL

The complexity of the inelastic response of RC structures may be significantly increased due to the interactive effects of applied arbitrary static and/or dynamic external loads. To simplify the analytical modeling of the inelastic behaviour of an isolated RC element, in most of the previous studies, the element axial forces have been considered as constant. However, during dynamic structural response under intensive earthquake ground motion, axial forces may be produced in some constituent structural elements with high variation around the initial static axial force levels. To improve the analytical simulation of structural static and/or dynamic inelastic response with present variation of element axial loads, the authors have developed and implemented in this study the so-called NON-MIFE (Nonlinear Multi-Interface Finite Element) model. With implementation of the developed NON-MIFE model of isolated arbitrary in-plane spaced structural elements (Fig. 4), formulation of nonlinear predictive model of the integral structural system has been made possible by applying the standard finite element and numerical analysis procedures.

The most essential steps in the present formulation of the nonlinear finite element (NON-MIFE) of an RC structural member exposed to interactive effects of reversed shear and varying compressive axial force may be summarized in the following: (1) A finite element is defined by two end nodal points and consists of an arbitrary number of sub-elements along the element length, (2) Each sub-element is defined by the so-called interface elements at its both ends, (3) Element local degrees of freedom consist of two end rotations along with axial deformation which are respectively related to the bending moments at both ends and the element axial force, (4) Two local degrees of freedom are defined for each interface-element (section) and include time variation of section curvature and section current axial strain, related respectively to the interface element bending moment and the current axial force, (5) The incremental local degrees of freedom and the forces of the interface element are appropriately related to local incremental deformations and forces at both element ends, (6) In the implemented step-by-step computational procedure for each interface element, axial stiffness and incremental axial strain is computed based on current increments of axial force and analytically simulated stress-strain relations for section materials, while section rotational stiffness is directly controlled by the adopted and predicted set of moment-curvature relations which provide incorporation of varying section inelastic hysteretic response depending on the current axial force level.

With the defined axial and rotational stiffness of each interface element along the considered single RC member satisfactory conditions are provided for simulation of the development and spreading of element nonlinearity depending on the current levels of fluctuating moments and time-varying axial forces, and the further procedure for computation of nonlinear stiffness matrix of RC finite element includes the following steps: (1) Computation of interface element local flexibility matrix by inversion of its previously computed stiffness matrix, (2) Computation of the element local flexibility matrix by integration along the element length, (3) Inversion of element local flexibility matrix to obtain element local stiffness matrix relating the current incremental element forces and deformations, and then further steps directly follow the standard finite element and matrix analysis procedure (Ref. 1).

PRACTICAL APPLICABILITY OF THE FORMULATED ANALYTICAL MODEL

It is necessary to point out some aspects related to the practical applicability of the formulated nonlinear multi-interface finite element (NON-MIFE) model for modeling of real structures composed of different types of RC members. If a structural RC member is rather slender, the input parameters of the envelope curve controlling the hysteretic moment-curvature relation of the structural member may be computed considering the specific level of axial force and dominant bending deformations. However, if the RC structural member is significantly exposed to shear deformations, the envelope curve for analytical simulation of the moment-curvature relation may be appropriately corrected to provide realistic simulation of the increased member deflection due to the existing shear deformations. To define the required parameters for this correction, some additional calculations based on a given member geometry and reinforcement distribution have to be made considering appropriately also the specific mechanical characteristics of the concrete and steel material used.

ANALYSIS OF SPECIMENS SHEAR FORCE - DEFLECTION ENVELOPE CURVES

To demonstrate the applicability of the proposed NON-MIFE model for analytical prediction of the integral structures inelastic response, the load-deflection envelope curve of the tested six-storey RC coupled shear wall system has been first analyzed. The formulated nonlinear mathematical model of this structure is shown in Fig. 5. It includes in total 24 nodal points and 28 finite elements. The first 11 finite elements are modelled as linear, where elements 2 to 11 are used to simulate the rigid zones to which the coupling beams are connected. The remaining 17 finite elements are modelled as nonlinear, where elements 12 to 23 are used to model the inelastic response of the left and right shear walls, and elements 24 to 28 actually simulate the nonlinear behaviour of the existing RC coupling beams.

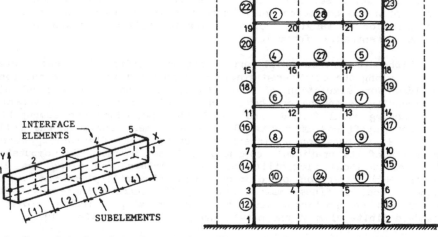

Figure 4. Formulated and implemented NON-MIFE model

Figure 5. Formulated nonlinear model of coupled shear wall specimen (CSWS)

In the analysis of the load deflection envelope curve of the modelled specimen, loading conditions have been completely simulated in the same way as in the previously conducted experimental test. The specimen was exposed to three simultaneous loadings, where the first two loading functions represent the axial loads acting in vertical Y-direction, denoted as N1(t) and N2(t) and applied in nodal points 23 and 24, respectively. The third loading function P(t) is used to simulate the applied shear loading in nodal point 23 in X-direction. In this analysis, the following loading conditions have been satisfied simulating realistically loading of the specimen in the previously conducted experimental test.

In the first phase, both shear walls were exposed to constant axial forces N1 = N2 = 44,000 kg, while in the second phase the specimen was step-wise exposed to increasing shear force P(t) and to simultaneously changing axial forces N1(t) and N2(t). Since the specimen was displaced by force P(t) to the right (positive X-direction), axial force N1(t) was gradually decreased, while axial force N2(t) was accordingly increased simulating in this way possible time-variation of axial forces under increasing horizontal displacements of the structure due to strong earthquake ground motion. Since reached minimum and maximum values of axial forces N1 and N2 were 21,700 kg and 66,000 kg, the maximum variation of axial forces reached a high amount of 102.7% and 50.0%, respectively. Under these complex loading conditions, this computation generally showed good correlation between analytically predicted and experimentally recorded nonlinear load-deflection envelope curve of the integral system (Fig. 6).

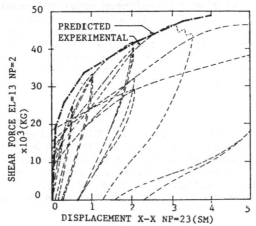

Figure 6. Comparison of theoretically predicted and experimentally recorded shear force - displacement envelope curve

Because the NON-MIFE model, as formulated and verified in this study, provides conditions for simulation of nonlinear behaviour of RC members under reversed cyclic loads, with this model has been newly made extension of the computer program NORA originally developed by the second author to carry out general nonlinear static and nonlinear dynamic analysis of integral reinforced concrete structures (Refs. 3 and 5).

MODELING AND INELASTIC EARTHQUAKE RESPONSE ANALYSIS OF THE TESTED SPECIMEN WITH THE PROPOSED MODEL AND DEVELOPED COMPUTER PROGRAM NORA

Since the assembled stiffness matrix of RC member, based on the developed NON-MIFE model, is time dependent, for multi-degree-of-freedom system earthquake response is computed by step-by-step solution of dynamic equilibrium equation written in incremental form (Eq. 1) considering constant structural stiffness for a small time interval Δt (Refs. 1 and 2).

$$[M]^{t+\Delta t}\{\ddot{u}\}+[c]^{t+\Delta t}\{\dot{u}\}+{}^{t}[k]\{u\} = {}^{t+\Delta t}\{R\} - {}^{t}\{F\} \qquad (1)$$

where [M] and [C] are mass and Rayleigh damping matrix, ${}^{t}[K]$ and ${}^{t}\{F\}$ stiffness matrix and vector of nodal point forces corresponding to element stresses at time Δt, ${}^{t+\Delta t}\{\ddot{u}\}$, ${}^{t+\Delta t}\{\dot{u}\}$ and ${}^{t+\Delta t}\{R\}$, vectors of nodal point accelerations, velocities and external forces for time t+Δt, and {u} displacement increments between time t and t+Δt. In the computer program NORA extended in this study, both Newmark's and Wilson's step-by-step numerical integration schemes have been employed to compute inelastic structural response discretized by the presently developed (NON-MIFE) finite element. Generally, some iterations are required for developed unbalanced nodal point forces during a small time increment to obtain an improved solution. However, in the following demonstrative numerical example of RC coupled shear wall structure, assembling of new structural stiffness matrix has been considered at every time step (t=0.005 sec) and no additional iterations have been carried out.

As shown in Fig. 5, the discrete nonlinear model of example RC tested coupled shear wall structure, 4.47 m in height, consists of 24 nodal points, 17 NON-MIFE finite elements which are further discretized by 5 sub-elements and 6 interface elements for which inelastic properties are included considering the actual steel and concrete section areas. Assuming fixed boundary conditions at nodes 1 and 2, the present structure model includes in total 60 degrees of freedom in the defined global X-Y coordinate system.

To demonstrate the capability of the present NON-MIFE model for inelastic earthquake response analysis of RC structures considering dynamically varying axial forces, example nonlinear analyses have been performed using real earthquake ground motion with intensity level EQI=2. This means that the implemented Ulcinj-Albatros earthquake record obtained during Montenegro-Yugoslavia Earthquake of April 15, 1979 with PGA=0.17 g is amplified by factor 2, so the PGA implemented in this analysis was made PGA = 0.34 g.

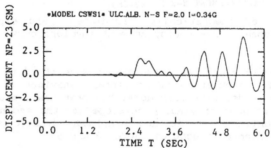

Figure 7. Predicted earthquake displacement history of NP=23 X-X direction

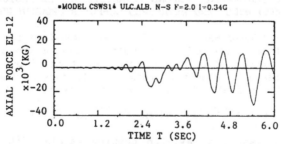

Figure 8. Computed dynamic time-variation of axial force in left shear wall EL=12, NP=1

Assuming existing element and additional nodal point masses, calculated initial dynamic characteristics, i.e., periods for the first two modes are obtained: T1=0.186 s and T2=0.054 s. In the present dynamic analysis, structural earthquake response is computed only for the first 6.0 sec, which actually include the most effective part of the considered input ground motion. It should be made clear that in the present formulation of the nonlinear structural model for earthquake response analysis, nodal point additional masses are specified considerably large to produce intensive nonlinear response; otherwise, due to the existing very large specimen stiffness, its dynamic response for the considered earthquake intensity will remain purely linear. In Figs. 7 through 10, selected results of the computed earthquake response are presented, demonstrating the most essential features of the model applicability to realistically simulate the earthquake intensity effects to the general structural nonlinear behaviour. Fig. 7 presents the displacement response of nodal point 23 in horizontal direction X-X with clearly expressed tendency of increased period of vibration due to significant softening of the structure as a result of induced nonlinearity. In Fig. 8, clearly shown is the produced large dynamic variation of the axial force in EL = 12 (around the level of the existing static axial force (N1 = 44,000 kg), as a result of intensive structural vibration. This variation is considered in the present analysis. Consequently, in Fig. 9, plotted are the predicted shear force - displacement hysteretic responses of the left shear wall EL = 12, NP = 1(a) and the right shear wall EL 13, NP = 2(b) with well expressed differences as a result of the existing dynamic variation of the axial forces. Finally, Fig. 10 shows the predicted moment-curvature hysteretic response of the right shear wall section EL = 13, NP = 2(a) and coupling beam EL = 26, NP = 13(b). It is clear that due to variation of the axial force, the moment capacity for the wall section is different for opposite sides, while for the case of coupling beam, moment capacity at both sides remains nearly constant, since insignificant dynamic variation of the axial force was produced.

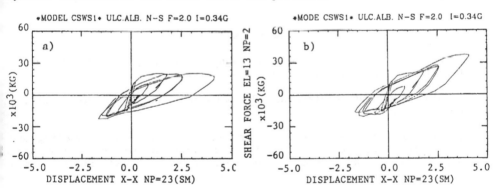

Figure 9. Predicted force - displacement hysteretic response of the left and right shear wall: EL=12, NP=1 (a) and EL=13, NP=2 (b)

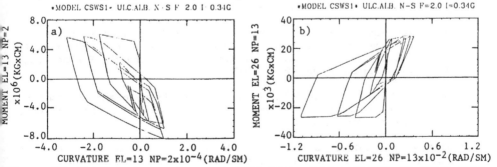

Figure 10. Predicted moment - curvature hysteretic response of shear wall and coupling beam section: EL=13, NP=2 (a) and EL=26, NP=13 (b)

CONCLUSIONS

In this paper, an applicable analytical modeling concept is presented for inelastic static and inelastic earthquake response analysis of integral RC structures, considering in the analysis the dynamically induced time-varying axial forces, and the following conclusions are obtained: (1) The newly formulated NON-MIFE (nonlinear multi-interface finite element) model possesses capability for realistic simulation of the inelastic response of RC members exposed to interactive effects of reversed bending and varying axial forces, as shown in the comparative presentation of the predicted and experimentally obtained results; (2) In the case of members with significant effects of shear deformations (as RC structural shear walls), the basic input envelope (M-ϕ) curves should be adopted with included effects of both shear and bending deformations; (3) dynamically induced highly varying axial forces have non-negligible effects on M-ϕ relations of structural sections and the seismic response of the integral RC structures; (4) For some specific types of engineering structures, high dynamic variation of the axial forces may be induced, as evident in the present example; and (5) the developed nonlinear model, the analytical procedure and the computer program NORA may be applied as sophisticated tool to carry out rigirous inelastic static and inelastic earthquake response analysis of real RC structures where consideration of the interactive effects of bending and time-varying axial loads have to be considered in order to realistically simulate the inelastic behaviour of the integral structure.

REFERENCES

1.   Bathe, K.J., Finite Element Procedures in Engineering Analyses, Prentice-Hall, Englewood Cliffs, N.Y., 1982.

2.   Clough, R.W. and Penzien, J., Dynamics of Structures, McGraw-Hill, New York, 1975.

3.   Ristić, D., Nonlinear Behaviour and Stress-Strain Based Modeling of Reinforced Concrete Structures under Earthquake Induced Bending and Varying Axial Loads, Ph.D. Dissertation, School of Civil Engineering, Kyoto University, Kyoto, Japan, June 1988.

4.   Petrovski, J. and Ristić, D., Nonlinear tests of bridge pier elements. Proceedings of the 8th European Conference in Earthquake Engineering, Vol. 4, Lisboa, 1986.

5.   Ristić, D., Yamada, Y., and Iemura, H., Inelastic stress-strain based seismic response prediction of RC structures considering dynamically varying axial forces. 9th World Conference in Earthquake Engineering, Tokyo-Kyoto, Japan, August 2-9, 1988.

6.   Simeonov, B., Experimental investigation of strength and deformation of reinforced concrete structural wall, Proceedings of the VII ECEE, Vol. 4, Athens, September 1982.

7.   Simeonov, B., Linear and nonlinear behaviour of reinforced concrete structural walls in multi-storey buildings, Ph.D. Dissertation, Civil Engineering Faculty, Belgrade, 1982 (original in Serbo-Croatian).

NONLINEAR RESPONSE OF ASYMMETRIC BUILDING STRUCTURES AND SEISMIC CODES: A STATE OF THE ART REVIEW

AVIGDOR RUTENBERG
Faculty of Civil Engineering
Technion - Israel Institute of Technology, Haifa, Israel

ABSTRACT

Analytical studies on the nonlinear earthquake response of single storey asymmetric structural models since the early 1960's are reviewed, with particular emphasis on structures designed by the static provisions of seismic codes. The effects of the main system parameters on peak ductility demand of the resisting elements and the maximum lateral floor displacement are discussed, including conflicting conclusions reported in the literature, mainly due to model dependence. The relative merits of several modern code provisions are compared. A new system definition is proposed to eliminate the model dependence of the results.

INTRODUCTION

For many years now it has been recognized that buildings with asymmetric configuration are more vulnerable to earthquake hazards than their symmetric counterparts. The recognition of this sensitivity led to the development of special code provisions to consider the additional forces which are likely to arise due to the amplification of the "static" eccentricity e between the mass center CM and the center of rigidity CR of the structural system (Fig. 1).

Perhaps the earliest of these provisions is the Mexico City code of 1959 (see [31]). It is not surprising that earthquake damage in asymmetric structures is expected to be controlled by design based on linear elastic analysis, usually equivalent static loading, since this is the approach to the seismic design of most symmetric structures.

The static seismic provisions of building codes are based on the traditional method in which the inertia forces are applied statically to the structure at the mass center (CM). For asymmetric structures this method takes the well known form (Fig. 1):

$$S_i = S \frac{k_i}{\Sigma k} \pm Se \frac{k_i a_i}{\Sigma ka^2} \tag{1}$$

in which:
S = static seismic shear on building specified by code
S_i = shear on element i
k_i = lateral stiffness of element i
e = eccentricity of S from CR (center of rigidity)
a_i = distance of element i from CR
PC = plastic centroid

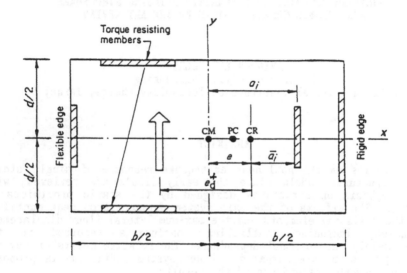

Figure 1. Plan of single storey monosymmetric model.

After the 1957 Mexico earthquake it became apparent that the static method is too simplistic due to its inability to predict the torsional effects with sufficient accuracy [31,32], and due to the large discrepancies between static and dynamic results [25]. These discrepancies occur because lateral-torsional coupling due to rotatory inertia of the floor mass modifies the eccentricity of the inertia forces (Fig. 1). The static methods were "saved" by the introduction of the design eccentricity concept. Calibration with linear dynamic analysis, mainly by means of the response spectrum approach, led to the amplification of the static eccentricity for members located on the flexible side of the floor deck, namely the side of CM away from CR, and to the reduction of eccentricity for members located near the rigid side (Fig. 1). Also, since it was shown that for a class of regular multistorey building structures a single storey model can adequately represent the linear torsional response (e.g. [19]), the calibration effort has mainly been carried out on such simple models.

The design eccentricity e_d replacing e in equation 1 usually takes the form:

$$e_d = \alpha e + \beta b \qquad (2)$$

in which b = plan dimension of building perpendicular to the direction of excitation. The coefficient α is an eccentricity modifier which is calibrated so as to obtain reasonable agreement between the equivalent

static analysis, as specified by the code, and dynamic analysis. The
coefficient β accounts for the effect of accidental eccentricity (for
review of likely sources see [32]). Numerical values for a and β as
specified by several seismic codes are given in Table 1. The rationale
behind the ATC-3 [C1] and the NEHRP [C6] disregard for the amplification of
eccentricity in dynamic analyses is discussed subsequently.

TABLE 1
Design eccentricities per several codes

code		e_d^+	e_d^-
ATC/1978	[C1]	e+0.05b	e-0.05b
NBCC/1985	[C2]	1.5e+0.10b	0.5e-0.10b
CEB/1987	[C3]	1.5e+0.05b	e-0.05b
SEAOC/1975	[C5]	e+0.05b	e-0.05b (e<0.05b)
UBC/1979	[C8]		0 (e>0.05b)
Mexico/1976	[18]	1.5e+0.10b	e-0.10b

Most of the calibration effort, however, focused on the response of
members located on the flexible side of the floor deck (Fig. 1) and less
attention was given to members on the rigid side, and to torsion resisting
members. Indeed, linear dynamic time history analyses have shown that the
static code provisions underestimate the response of members on the rigid
side [36]. It was expected that in strong earthquakes this would lead to
higher ductility demands on these members, as will be shown later. Some
typical linear results are given in Fig. 2 [37]. Figure 2 compares
normalized (asymmetric/symmetric) maximum edge displacements $y_{i,max}/y_{o,max}$
from an ensemble of time histories with results using several e_d formulas
(plots denoted eqns. 16 and 17 in Fig. 2b are proposals given in [37]).

NONLINEAR BEHAVIOUR AND DUCTILITY DEMAND: EARLIER STUDIES

For economic reasons most structures are designed to yield under severe
seismic loads. It is convenient to approximate their behaviour by means of
elastic-plastic or bilinear force-displacement relationship. The ductility
demand (y_{max}/y_y in Fig. 3) is a measure of the inelastic deformation, and
traditionally has been used as a damage parameter. It is not a very
satisfactory measure of damage because it ignores strength and stiffness
deterioration, pinching due to crack closing and the sequential nature of
plastic hinge formation. Moreover, studies using more realistic cyclic
force-displacement relations have shown that the ductility demand of
degrading models is similar to that obtained from bilinear models (e.g.,
[13, 48]). This is the main reason why many investigators prefer to use the
latter model. Because a number of elements is involved, peak ductility
demand (PDD) - the maximum demand of all the resisting elements - is the
parameter of interest.

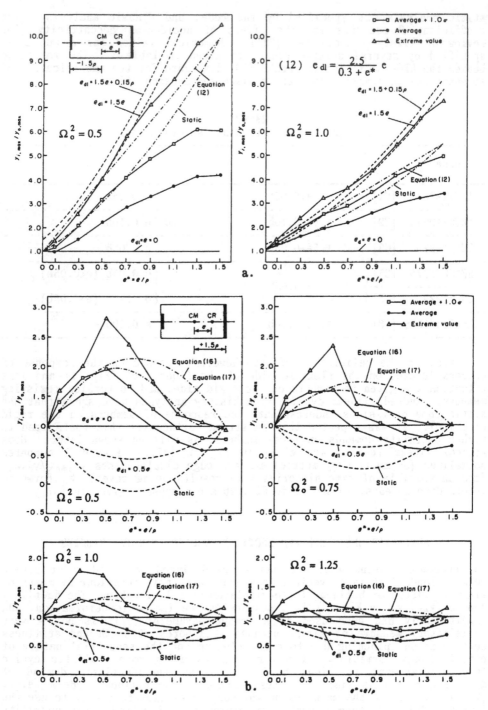

Figure 2. Typical linear results [37].
a. Flexible edge. b. Rigid edge.

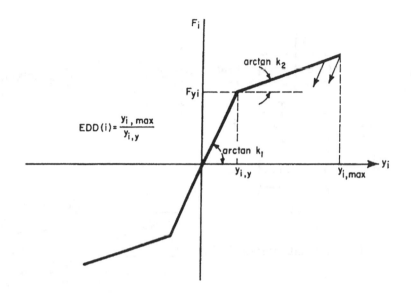

Figure 3. Bilinear force-displacement relationship.

Different categories of structures have different levels of ductility supply or capacity. The static approach to seismic design deals with these by assigning to them different values of strength reduction factors (R). The more ductile the structure the less design strength it needs. Brittle failure is assigned a value of 1.0 (strength based on elastic analysis), provided no overstrength is available, and highly ductile behaviour a much higher value: 4.0 to 7.0, depending on the code.

The study of nonlinear response of asymmetric structures began in the early 1960's. The earliest study appears to be that of Tanabashi et al. [45] who analyzed a bilinear 4-frame system under sinusoidal excitation using an analogue computer. The columns were excited in biaxial bending, but no post-yield interaction was considered.

Studies of similar nature were carried out later by Shiga et al. [41,42]. Considering the computing difficulties in that decade, these studies could not be very comprehensive, yet they showed that the strength redistribution among resisting elements affects ductility demand and reduces damage due to torsion. A later study by Erdik [16] dealt mainly with elastic-perfectly plastic displacement and rotation response spectra of mass eccentric yet structurally symmetric systems having four and sixteen identical elements with stiffness in two directions (Fig. 4). Erdik concluded that for small eccentricities these systems behave like single degree of freedom oscillators which, after yielding, respond mainly in translation. He also concluded that the displacements of the shear center are similar to those of multilinear systems with a flat top. Regarding the amplification of eccentricity due to modal coupling he found that this effect is still present, but tends to fade with increasing nonlinearity.

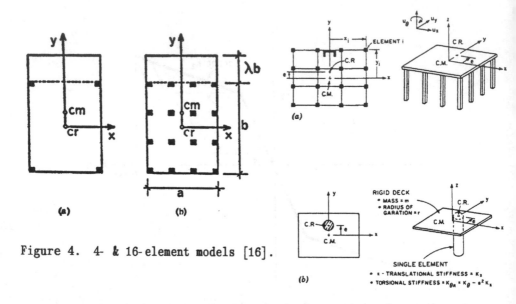

Figure 4. 4- & 16-element models [16].

Figure 5. Equivalent 1-element model [28,29].

A study on the ductility demand of simple five storey building structures having elastic-plastic force-displacement relationship and with either rigid or flexible floor diaphragms was made by Cardona [10]. He concluded that the ductility demand (DD) of eccentric flexible diaphragm buildings is higher than that of rigid diaphragm ones, and that the maximum demand occurs when e/b ≈ 0.3 (b = width of building normal to the direction of excitation, Fig. 1). He also found that the DD of long period eccentric systems with rigid floor diaphragms is lower than that of similar but symmetric systems.

In the late 1970's and early 1980's several studies were carried out in order to shed more light on the behaviour of yielding asymmetric systems. Kan and Chopra [28,29] assumed elastic-perfectly plastic behaviour, and approximated the multi-element system by a single element having a circular shear-torque yield surface using a single time history (El Centro). They varied the eccentricity ratio $e^* = e/\rho$ (ρ = mass radius of gyration), Ω (torsional to lateral frequency ratio), and the lateral vibration period T_0. They still found that amplification due to lateral-torsional coupling in the elastic-plastic range depends on Ω and as in elastic systems, being most prominent at $\Omega = 1$, but concluded, as Erdik [16] did, that it affects the maximum deformations in inelastic systems less than in linear ones. They also found that the vector sum corner displacement at the flexible edge column (Fig. 5) may reach three times the value at the mass center, which is much more than in the elastic case. However, this occurred within a narrow range of periods, suggesting to the authors of that study the presence of lateral-torsional amplification.

It is interesting to note that several years later a single element model was also used by Palazzo and Fraternali [30] who concluded that amplification of DD occurs at $\Omega = 1.0$.

The effects of plastic X-Y interaction on the response to the El Centro record of 4-element elastic-perfectly plastic systems of the type shown in Fig. 4 were studied by Yamazaki [51], whose variables were e*, T_0, and excitation level. He concluded that the effects of X-Y interaction on maximum displacements are not significant. Yet, interaction tends to smooth the peaks observed in the response of similar, but not interactive, elastic-plastic systems. Yamazaki also studied the effects of bidirectional input, eccentricity in two directions, and natural period. He observed that translational displacements are not very sensitive to eccentricity, particularly in longer period systems, and although torsional response increases with eccentricity it is more pronounced in short period systems. Finally he found that lateral and torsional displacements fall with increasing strength level.

In most of the studies described so far interest focused on maximum displacement and rotations, but ductility demand was not considered explicitly. This is perhaps understandable in view of the fact that in all the elements of the models used strength was proportional to elastic stiffness, so that although absolute ductility levels were not computed the ductility demand of the elements was proportional to their larger absolute lateral displacement (ignoring the interaction effect). The study of Irvine and Kountouris [26,27] is mainly concerned with the ductility demand of asymmetric systems as compared with that of similar but symmetric ones. Their analyses used two equal elastic-plastic elements as the lateral load resisting system (Fig. 6) - a system which is statically determinate. Based on several time histories the following conclusions were reached:

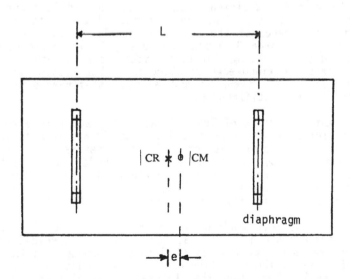

Figure 6. 2-element model [26,27]

* Peak ductility demand (PDD) is insensitive to e* and to Ω.
* PDD is seldom more than 30 % larger than in a similar but symmetric system.
* The element with the PDD is the one closer to the mass center (i.e., near the flexible edge)

The reader may have already noticed differences in the conclusions of some studies, suggesting that the problem is quite complex. The discussion of these and other discrepancies is made in the following section.

In the following years other investigators started to publish their findings. In a series of papers Tso and his coworkers [5,46-48] reported the results of parametric studies on the bilinear response of systems of the types shown in Fig. 7b. and 7c. The PDD and the maximum edge displacement were chosen as the main response parameters. The effect of bidirectional interaction on yielding was studied by Tso and Sadek [47]. They found that interaction lowered the response to some extent but on the whole X-Y interaction was not significant. Similar results were reported by Shohet [43]. Tso and Sadek [47] also concluded that the PDD of systems with small eccentricity is similar to that of similar symmetric ones, but PDD is much larger for large eccentricities. Tso and Sadek [48] and Bozorgnia and Tso [5] reported that lateral-torsional amplification does not occur in the yielding systems since the structure is sufficiently detuned, and that the response is primarily not in translation, results which are at some variance with earlier findings by Kan and Chopra. They also found that the effect of asymmetry on PDD and maximum edge displacement was the strongest for rigid systems at high strength reduction factors. The torsional to lateral frequency ratio was not found to affect the PDD significantly. The characteristics of the input motion i.e. the frequency content and the presence of long pulses were found to be of some importance. They also concluded that the stiffness eccentricity e amplifies the response, and that the effect is strongest for low period systems. When comparing the ductility demand of the 3-element model of Tso and Sadek [48] with the 2-element model of Irvine and Kountouris [26,27] Tso and Sadek found that the DD of the latter model was on the whole lower. Thus the model dependence of the nonlinear results became quite evident. In another study on three 3-element systems [46] it was concluded that systems with elements having identical yield displacements but uneven strengths had larger maximum edge displacements than systems with elements having either identical strengths and elastic stiffnesses or identical strengths but different stiffnesses.

An important conclusion of Tso and Sadek [47] for future research was that degrading force-displacement relationships did not lead to larger PDD, suggesting that elastic-plastic or bilinear systems with very small secondary slope ratio which do not faithfully model reinforced concrete behaviour can nevertheless predict adequately its PDD. However, this conclusion cannot be extended to X-braced frames for which it was found that the bilinear model significantly underestimates PDD [7].

Syamal and Pekau [44] studied the inelastic response of single storey monosymmetric systems shown in Fig. 8, subjected to sinusoidal excitation. In their model eccentricity was effected through changes in the stiffness of the two elements in the direction of excitation, but their yield levels were taken as equal. They found that lateral-torsional coupling was not an important parameter and concluded that PDD is strongly affected by stiffness eccentricity and is the largest in torsionally flexible systems (Fig. 9).

Bruneau and Mahin [6-9] reported the results of parametric studies on the bilinear response of monosymmetric single storey systems, including some with two elements. On the basis of a particular 2-element model they proposed a 1-DoF (degree of freedom) model from which the ductility demand

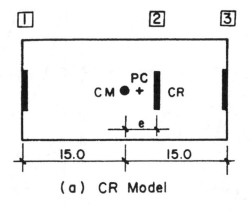

(a) CR Model

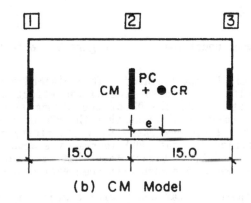

(b) CM Model

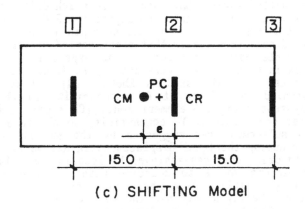

(c) SHIFTING Model

Figure 7. 3-element models.

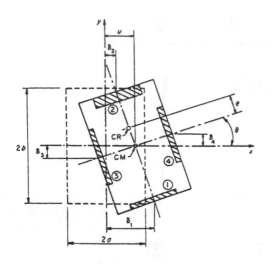

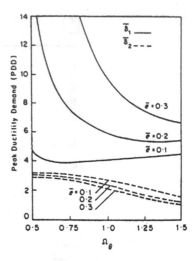

Figure 8. 4-element model [44].   Figure 9. Effect of Ω on PDD [44]

of 2-DoF systems could be approximated. They claimed reasonable agreement
with the exact results. They also studied the effect of PC eccentricity
and concluded that it affects the DD of the strong element (stiff edge).
Also interesting are their conclusions that the PDD of these systems is not
very sensitive to the frequency ratio Ω, eccentricity ratio e*, and the
uncoupled natural period T_0, that PDD increases somewhat with Ω (!), and
that DD of 2-element models is higher than that of 4-element ones. These
last two conclusions are at variance with some earlier results.

Many of the studies so far dealt with systems in which strength was
proportional to stiffness, i.e. the plastic centroid PC - or the center of
resistance at yield - coincided with CR (Fig. 1). These are not realistic
systems in the sense that in code-designed structures strength is allocated
by means of equation 1 (with e_d replacing e). Thus this proportionality is
lost, and PC is shifted towards CM. Studies in which different strength to
stiffness ratios were assumed started to appear in 1986. Tso and Bozorgnia
[46] concluded that systems with constant strength to stiffness ratio had
the largest edge displacement, also that short period systems are likely to
have larger flexible edge displacement - an expected result. However,
ductility demands in these systems were not reported, nor was the effect of
PC location. These were discussed by Shohet [43] and Rutenberg et al.
[35]. They studied the effect of PC eccentricity on the response of
monosymmetric systems using a 3-element model (CR model in Fig. 7). The
idea of exploring the effect of PC eccentricity to obtain a lower PDD and
lower maximum displacement was suggested by the success in eliminating
twist and thus reducing the total displacement in asymmetric base isolated
multistorey buildings. This was achieved by locating the plastic centroid
of the base isolators so as to coincide with the mass center of the
superstructure (e.g., [15]). From their parametric study they concluded
that the member on the rigid edge was the most vulnerable with respect to
ductility and that the flexible edge displaced the most. They pointed out
that the need to increase the strength of element 3 (Fig. 7) required by

the shift of PC towards CR with falling frequency ratio is in agreement with earlier linear and nonlinear results [36,37,44]. They also found that DD can be controlled by strength redistribution among the resisting elements, and that the best design strategy for minimizing PDD was that for which PC is located approximately halfway between CM and CR. A similar conclusion was obtained by Yiaslas et al. [50]. Rutenberg and his coworkers also concluded that PDD falls with increasing Ω, which is at some variance with Bozorgnia and Tso [5].

Sadek and Tso [40] also studies the effect of PC eccentricity on DD and maximum displacement. Using a 4-element model with identical properties in the x and y directions and located symmetrically about CM, they also concluded that PC eccentricity is a parameter that should be considered in the study of seismically excited yielding asymmetric structures. They appear to have been concerned mainly with elements on the flexible side of CR. These studies still did not address code-designed structures, since strength was allocated in some arbitrary fashion rather than through equation 1.

Yet it became apparent that the strength distribution among the elements, and PC eccentricity which gives some measure of this distribution, affect PDD and maximum displacements of realistic asymmetric systems.

Goel and Chopra [20,21] studied the effects of total strength and its distribution as determined by the number and location of the resisting elements, as well as that of PC eccentricity on the inelastic response of single storey systems. A new parameter was introduced, namely the effect of elements oriented normal to the direction of excitation. The ground motions used were a half-cycle displacement pulse and the El Centro record. Some important conclusions are summarized below:

* The normal elements reduce the torsional deformation, although their effect on PDD was not significant except perhaps in the very short period range (half cycle excitation). However, the significance of PDD values on the order of several hundreds is not clear.
* The effect of number of elements in the direction of load on displacements was not significant, although they affect PDD (again 2-element systems have lower DD than 3-element ones).
* PDD of mass eccentric systems (shifting model) is likely to be higher than stiffness eccentric ones (CM & CR models), particularly when $e_{pl} \Rightarrow 0$. This conclusion refers to the $\Omega = 1.0$ case. Note that this effect was found by the author (unpublished results) to be much more prominent for $\Omega = 0.8$ than for $\Omega = 1.0$.
* Element deformation (y_{max}) is close to that of symmetric systems when the strength reduction factor R is larger.
* Period dependence of the response is quite low and decreases with excitation level.

Their conclusion regarding code-designed systems are discussed subsequently.

In the mid-1980's investigators began to realize that simple strength allocation rules, such as equal yield displacement and some arbitrary variations thereto designed to bring about shifts in PC location, are unlikely to shed light on the ductility demand of code-designed structures, since the strength of their resisting elements is allocated on the basis of equation 1 (with e_d replacing e).

CODE- DESIGNED STRUCTURES

Perhaps the earliest work that attempted to study code designed structures was that of Shohet [43] (see also [35,39]). He studied the response of the 3- element model CR in Fig. 7 designed according to the NBCC [C2] and the ATC [C1] provisions using several earthquake time histories. He concluded that also for these systems the element located near the stiff edge is the most vulnerable, because its DD was the highest among the three. Also that the Canadian code designs performed better than the ATC designs because the NBCC allocated relatively more strength to the stiff edge element, and because the Canadian designs had higher strength than the ATC ones. As noted, the need to allot more strength to the stiff edge elements was more important for torsionally flexible systems. The peak displacements at the flexible edge were not appreciably affected by strength redistribution. The study also confirmed earlier findings: PDD is affected by the energy content of the earthquake at periods higher than the structural, and that there is no amplification at $\Omega = 1.0$. He noted that in order to overcome model dependence additional parameters are needed, and suggested rotational strength as one.

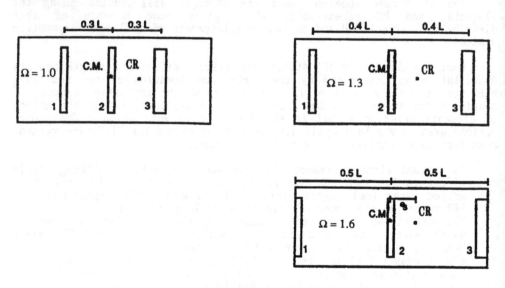

Figure 10.   3- element models [13].

The study of Diaz- Molina [13] also dealt with the response of code-designed asymmetric single storey structures, using models of the type shown in Fig. 10. He came to similar conclusions as Rutenberg and his coworkers, namely that the elements at the flexible edge displaced the most and those at the stiff edge showed the largest PDD. Therefore, to lower their DD it is advantageous to use a reduced value of e_d for these elements, as in the NBCC [C2] where $e_d = 0.5e$, or to ignore torsional effects as in the SEAOC code [C8] or the UBC [C5]. However, he did not consider the effect of overstrength due to the amplification and reduction of e_d relative to e. Regarding the effect of Ω and e on the response he found that PDD falls with increasing Ω and decreasing e (Fig 11). He also

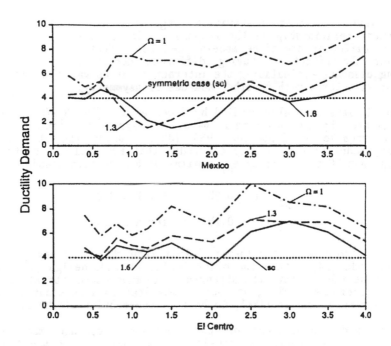

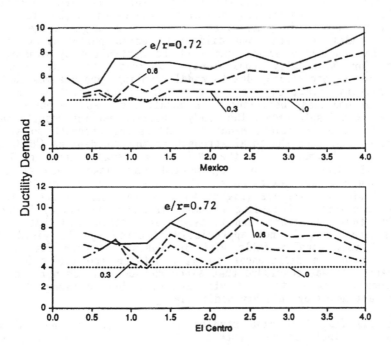

Figure 11.   Effect of Ω and e on PDD [13].

made a very important observation regarding the effect of force-displacement relationship on the response. It appears that not only the choice of different degrading models does not affect DD appreciably, but also that the effect of having different degrading rates among the resisting elements - simulating the interaction of RC walls and frames - is small for structures with no other sources of asymmetry.

Code-designs were also studied by Gomez et al. [22,23] (also [17]) who were motivated by the poor performance of asymmetric structures in the Mexico earthquake of September 1985 [33]. Using the shifting (mass eccentric) model of Fig. 7 they concluded that PDD could be controlled by requiring that PC eccentricity e_{pl} remains within certain bounds:

$$e_{pl} \geq e - 0.2b \qquad \text{for } R \leq 3.0$$
$$e_{pl} \geq e - 0.1b \qquad \text{for } R > 3.0 \tag{3}$$

where R = seismic behaviour factor. This requirement is now part of the Mexico DF code [24]. When the older Mexico DF code-design [18] does not satisfy these requirements the strength of elements near the stiff edge is to be duly increased. This, of course, may lead to a substantial increase in total strength, and is likely to be too conservative [11,20,21].

Since model dependence of the nonlinear response made it very difficult to draw general conclusions from single type model studies, Rutenberg et al. [39] extended the work of Shohet [43] by considering three models, thereby covering a wide range of possible structural configurations (Fig. 7). They compared the response of designs based on the ATC [C1] rules and those of NBCC [C2] by means of a relatively large parametric study. These two codes represent two different design philosophies: the NBCC assumes that ductility demand can best be controlled by dynamic linear analysis (or statically equivalent), whereas the ATC ignores the amplification of eccentritiy due to modal coupling, and assumes that static linear analysis (i.e., $e_d = e$ + accidental eccentricity) is an adequate predictor of nonlinear behaviour. As usual, the response parameters were PDD and maximum displacement. The study confirmed earlier results that $e_d = 0.5\ell - 0.1b$ is an efficient means for allocating strength to stiff edge elements, so that when the total strength of the two designs (NBCC and ATC) is made equal ("normalized"), the NBCC design usually leads to lower PDD than the ATC does, even for long period earthquakes such as the 1977 Bucharest and the 1985 Mexico ones. The beneficial effect of designing for higher rotational stiffness (higher Ω) also for the nonlinear range has also been confirmed. Regarding maximum displacements (y_{max}) it was again found that they are higher than in symmetric systems, and that for higher period systems y_{max} is not appreciably affected by the yield strength level, but y_{max} usually falls with increasing strength for low period systems. Also the differences in y_{max} between the two codes are not significant. It was concluded that since PDD and y_{max} do not occur in the same element, a more efficient strength distribution among the elements should be attempted by suitably modifying e_d.

Tso and Ying [49] also studied the ductility demand and y_{max} of monosymmetric single storey systems. Seven ways of specifying strength distribution among the elements were considered: strength proportional to stiffness ($e_{pl} = e$), static ($e_d = e$), and the following code rules: NBCC/1985 [C2], Mexico (1976 [18] + 1987 Complementary Technical Norms

[24]), New Zealand/1984 [C7] and UBC/1988 [C5]. The model used was of the CM type (Fig. 7). Comparisons of the resulting total overstrength due to $e_d \neq e$ and of the PC eccentricities were made, and two "generic" models were proposed which were believed to be representative of the various code designs in their total strength and in its distribution among the elements. Eight ground motion records of similar frequency content were applied. It appears that the study focused on the ductility demand of the flexible edge element, and no account was taken of the response of the rigid edge or central elements. A strong influence of the strength distribution on the response was again observed. Tso and Ying [49] concluded that the DD of the flexible edge element is lower when its strength is allocated through a code formula, namely with small e_{pl} compared with the $e_{pl}=e$ case, but y_{max} is higher. Regarding the new Mexico Complementary Technical Norms [24] they found that although large overstrength is required, improvement in DD was marginal. This conclusion is not inconsistent with other studies of the flexible edge element.

Goel and Chopra [20,21] also studied the response of code designed structures excited by a half-cycle displacement pulse and by the 1940 El Centro record. They are in agreement with earlier studies: PDD is to be found in stiff edge elements and y_{max} in flexible edge elements, NBCC or UBC type formula for stiff side e_d adequately controls PDD, and as noted, the 1987 Mexico Complementary Norms are unnecessary. They also observed, as earlier investigators, that suitable modifications in e_d can improve the response, and rightly pointed out that the linear and nonlinear requirements from the flexible side element may not be compatible.

Duan and Chandler [14] also studied 3-element models of the CM type (Fig. 7), and allotted strengths to elements by means of the Eurocode 8 formula [C4]. Again, PDD and y_{max} were the main response parameters. They found that the stiff edge element has the largest PDD, and observed that significant increase in PDD is to be expected in the proximity of the predominant input frequency. The effect of Ω on PDD was not found to be significant. However, the range of Ω studied was 0.7 → 1.0 and strength was not normalized, so that lower Ω may have led to somewhat higher total strength. A very recent paper of Chandler and Duan [12] concludes that $e_d = 0.5e - 0.1b$, as in the NBCC is to be preferred. They also concluded, as noted, that the 1987 Mexico Complementary Norms [24] regarding e_{pl} is overly conservative. An interesting observation made by these investigators regarding the inclusion of accidental eccentricity in the dynamic analysis will be discussed in the following section.

Benbenishti and Rutenberg [3,4] studied the effect of elements normal to the direction of excitation on the response of code designed asymmetric systems (Fig. 1). It appears that in many cases these elements are likely to lower the PDD. These results are consistent with Goel and Chopra [20,21] described earlier. The effect, however, depends on the type of excitation and there are period ranges where these effects are less noticeable. It is interesting to note that above a certain threshold an increase in their absolute strength has no additional effect. This matter deserves further study because these elements are practically always present.

DISCUSSION

The picture emerging from the foregoing review is somewhat confusing, and the main conclusion is that, in addition to the linear properties of the system and element stiffness and location, the overall strength and its distribution among the elements are the most important parameters affecting the peak ductility demand of bilinear asymmetric systems. On the other hand, maximum displacements are easier to predict since they are less sensitive to strength distribution.

It is quite clear that the parameters used to define the nonlinear systems considered in this paper, namely total yield strength and plastic centroidal eccentricity (e_{pl}) are insufficient for unique definition. Moreover, even the linear parameters are not uniformly defined by the investigators in the field. It appears that most of the problems and disagreements among researchers are the result of using different models (assuming, of course, no human error, which is bound to occur in such complex problems). This is not surprising considering the large number of parameters involved. In a recent paper Chandler and Duan [11] discussed some of these differences.

Already in the linear range there are different possibilities for defining the key parameter Ω and several choices for the damping ratio.

Frequency Ratio Ω. There are three possible ways of defining Ω [3]: (1) Ω_M - with respect to CM, (2) Ω_R - with respect to CR, and (3) Ω_0 - mass moment of inertia with respect to CM $(J_{\theta M})$, and torsional rigidity with respect to CR $(K_{\theta R})$. The following relations result:

$$\Omega_M^2 = \omega_{\theta 0}^2/\omega_y^2 = mK_{\theta R}/(J_{\theta M}\Sigma K_y)$$

$$\Omega_M^2 = \Omega_0^2 + e^{*2}; \qquad \Omega_R^2 = \Omega_0^2/(1+e^{*2}) \qquad (4)$$

$$e^* = e/\rho; \quad \rho = \sqrt{J_{\theta M}/m}$$

in which m = mass of deck, $\omega_y = \sqrt{\Sigma K_y/m}$ = lateral frequency of system for e = 0, $\omega_{\theta 0}^2 = K_{\theta R}/J_{\theta M}$. Evidently $\Omega_R < \Omega_0 < \Omega_M$, and the difference between Ω_R^2 and Ω_M^2 increases with increasing e*. Therefore two models having equal numerical values of Ω_M and Ω_R represent in fact two different structural systems. The advantage for parametric studies of using Ω_0 (which does not appear alone in the equations of motion) is quite clear: it is possible to vary e without changing any of the other physical parameters, i.e., the mass distribution and the structural system remain intact - it is only the mass center that is being shifted. Regarding the choice between Ω_M and Ω_R in the equations of motion it is a matter of convenience, and it is well known that for small Ω_M large eccentricities have no physical meaning [6,34]. For multistorey buildings it is often difficult to define a CR axis so that CM is a natural choice.

Damping ratio. Damping ratios from 2 to 5 percent have been used by different investigators. However, this does not appear to be a major source of variance.

The main difficulty, however, is the nonlinear range. Whereas it is possible to define linear systems by T_0 (or ω) Ω, e^* (and damping ratio), these parameters are insufficient for unique definition of yielding systems. As already noted, in most of the early studies described herein, it was assumed that member strength was proportional to stiffness i.e., strength/stiffness was identical for all resisting elements. From this assumption it follows that PC is located at CR. However, different systems, with identical PC location but different element location have different PDD. It has been noted that 2-element models have somewhat lower PDD than 3-element models [48]. Contrary results were also published [6], but Chandler and Duan [11] observed that different value of the basic parameters have led to the discrepancy. It was suggested [48] that the lower dependence of DD on e of the 2-element models is probably the result of the independence of element forces on the strength distribution in statically determinate system, in contradistinction to 3-element models. This may also lead to a different sequence of plastification. Also, for a given rotation the displacement of a 3-element model is larger than of 2-element one. The edge member in the former model is usually further away from the instantaneous center of rotation, and its yield displacement will be lower than that of the latter, leading to higher PDD. In fact the rotational strength of the 2-element model is higher so that the two systems are not similar in the nonlinear range.

With code-designed structures new problems arise. First, since e_d has two values: one for each side of the deck, the total strength of asymmetric systems is higher than that of their symmetric counterparts, again leading to different results if overstrength is not considered. Secondly, the strength to stiffness ratios of the elements are no longer constant and depend on their stiffness and location relative to CR. It is interesting to note that PC eccentricity in code-designed structures is usually much smaller than e, the actual value of e_{pl} depending on the particular code chosen and model used. Thirdly, some models include elements perpendicular to the direction of excitation, other models do not. Fourthly, most investigators use 3-element models, but the PDD of 2-element models and also of 4-element models has been shown to be different.

The inclusion of accidental eccentricity in evaluating e_d (equation 2) is controversial. Chandler and Duan [12] advocate excluding the accidental eccentricity (βb) from the expression. For a given code, the resulting total strength and its distribution (equation 1) depends on whether or not βb is included in e_d. Other definition difficulties are relevant to all types of models:

* Excitation level (or R value): Different investigators use different R values (some use 3, others use 4 or 5) but PDD(asymmetric)/PDD (symmetric) depends on R.
* Secondary slope ratio: Most investigators do not assume pure elastic-plastic response, they usually choose a bilinear force-displacement relationship with a mild secondary slope (some use 2% others use 3% or higher). PDD depends to some extent on the secondary slope ratio.

Finally, some investigators base their conclusions on the response to one earthquake record. However, the frequency content, the presence of long peaks and the different plastification sequences that may result require the study of results based on several time histories.

Goel and Chopra [20,21] observed that since the total strength of code designed asymmetric structures is higher than that of their symmetric counterparts, results of studies assuming normalized total strength (i.e., asymmetric strength = symmetric strength) are not directly applicable to these structures. Such a claim is not warranted since the differences among code-designs derive from two sources: (1) different total strength (2) different strength distribution. In order to study the effect of strength distribution on ductility demand it is necessary to neutralize total strength effects since higher overall strength lowers the overall ductility demand. Also, different codes have different design base shear coefficient c_d, yet these investigators advocate the use of a normalized value of c_d, which is not consistent with their observation about overstrength due to variations in e_d. Finally, higher overall strength results in a more expensive structural system. It therefore appears that it is important to study the behaviour of code designed asymmetric structures on equal strength basis if one is to make a meaningful comparison among the various codes and to suggest improvements in the e_d formula.

Goel and Chopra [20,21] also observed that since models without eccentrically located elements normal to the direction of excitation are less common, the results based on such models are not relevant to code designed structures. Evidently, the stiffness of edge elements (in the direction of excitation) of models with elements normal to the direction of excitation is lower than that of similarly located edge elements in models without the normal elements because the rotational stiffnesses of the two models must be equal to maintain the same value of Ω. Therefore their strengths (computed per eqn. 1 with e = e_d) is also lower, so that drastic variations in PDD due to the presence of the normal elements should not be expected. The valid part of the argument is that systems with normal elements can have larger rotational *strength* than models without them (although the two rotational *stiffnesses* are equal). Studies by Benbenishti and Rutenberg [3,4] indicated, as noted, that the gain is rather modest.

An interesting observation was made by Chandler and Duan [12] regarding the value of e_d to be used in equation 1. Since in standard analysis there are no uncertainties in specifying the element properties and mass distribution and no torsional component of the ground motion, the accidental eccentricity should not be included in e_d when allocating strength per equation 1. It follows that including βb in e_d is not conservative. Whereas this argument is evidently correct, the proposed approach will result in studying a model which is different from the one being considered (which includes βb).

The foregoing discussion has shown that the lack of a unique system definition by an accepted set of parameters is perhaps the main obstacle to real progress in this field. It is therefore important that a suitable model be agreed upon by investigators. This model should possess the main characteristics of typical asymmetric building structures (single and multistorey), yet it should also be sufficiently simple to permit extensive parametric studies. A suggestion towards this goal is made in the last section of this paper.

SUMMARY OF CONCLUSIONS: CODE-DESIGNED ASYMMETRIC STRUCTURES

Several discrepancies and inconsistencies among investigators have been reported in the preceding sections.   Yet some general conclusions about code designed structures do emerge from the studies reviewed in this paper (and from unpublished investigations by the author):

* The peak ductility demand (PDD) of asymmetric systems is larger than that of the corresponding symmetric systems, yet they follow the same trends: PDD falls with increasing total yield strength and for the California type records (e.g., El Centro, Taft) with increasing natural period.
* The PDD of systems designed by codes which allocate more strength to elements on the stiff side than required by statics, such as the NBCC/1985, SEAOC/1988 (or UBC/1988) is lower than that of designs based on ATC/1978 (or NEHRP/1988) type strength distribution which is mainly by statics (the accidental eccentricity modifies this somewhat).   Usually the most sensitive element is the one located near the stiff edge of the deck.   The larger overstrength of the NBCC and SEAOC/UBC designs compared with that of the ATC designs is only partly responsible for their superior behaviour (see next item).
* The total yield strength of code designed asymmetric systems is higher than their symmetric counterparts.   The overstrength increases with stiffness eccentricity and for a class of models may also increase with falling frequency ratio Ω.   The extent of overstrength depends on the coefficients a and β in equation 2.
* The eccentricity of the plastic centroid (PC) e_{pl} does not appear to be a useful parameter for allocating strength to the resisting members.   For code-designed structures e_{pl} is usually much smaller than the stiffness eccentricity e, but it is model dependent.
* The stiffness eccentricity e has some effect on PDD. This depends on the nature of excitation and the model. However, ATC designed shifting (mass eccentric) models appear to be more sensitive to variation in e than the CR and CM models (stiffness eccentric), particularly to the Mexico 1985 record when e becomes large (not shown).
* Torsional to lateral frequency Ω ratio has some effects on PDD.   It usually falls with increasing Ω when the total strength of the systems compared is made equal. This is important since the different results obtained by different investigators are probably due to different overall strengths (e.g., in some models total strength increases with Ω thus reversing the trend).
* There does not appear to be a clear dependence of PDD (asymmetric)/PDD (symmetric) on period provided R (strength reduction factor) is not small. Again, the shifting model appears to be the most sensitive.
* The ductility demand of systems with elements normal to the direction of excitation is likely to be lower than systems without these elements.   However, the reduction is not large. Note that increasing the strength of these elements above a certain threshold does not affect PDD.
* The effect on DD and y_{max} of nonlinear X-Y interaction in biaxially loaded elements does not appear to be significant.
* The elements located at the flexible edge of the deck displace the most, their displacements, however, are not much larger than those of symmetric systems.
* Maximum displacements usually increase with natural period (California records mainly).

* For low period systems displacements fall with increasing yield
strength as in the symmetric case.

PROPOSALS FOR FUTURE RESEARCH

The foregoing review has shown that there is some agreement about many
aspects of the seismic response of code designed asymmetric structures.
Yet, some important issues need to be addressed.

The main difficulty appears to be that of system definition: It was
shown that the nonlinear parameters used today, namely, excitation level,
total strength, PC eccentricity, number of elements, presence or absence of
normal elements are insufficient. A proposal to consider the total
rotational strength as an additional parameter [43] is a step in the right
direction, but unpublished studies by the author have shown that this is
still insufficient. What is missing, of course, is a simple definition of
the translational and of the rotational strength-stiffness (or force-
displacement and torque-rotation) relationships for the system. A
force-displacement plot of a typical system is shown in Fig. 12.

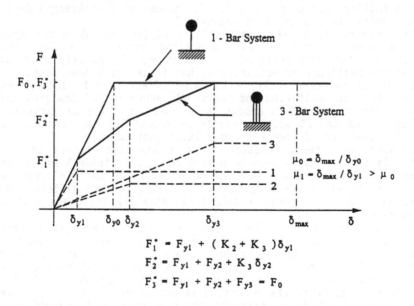

$$F_1^* = F_{y1} + (K_2 + K_3)\delta_{y1}$$

$$F_2^* = F_{y1} + F_{y2} + K_3 \delta_{y2}$$

$$F_3^* = F_{y1} + F_{y2} + F_{y3} = F_0$$

Figure 12. Actual multilinear force-displacement relationship
for a 3-element system [38].

Some preliminary results using this approach are given in Fig. 13
[38]. The Figure compares the results of the correct translational or
symmetric behaviour of the system, i.e., having a multilinear
force-displacement relation, with the asymmetric behaviour of the same

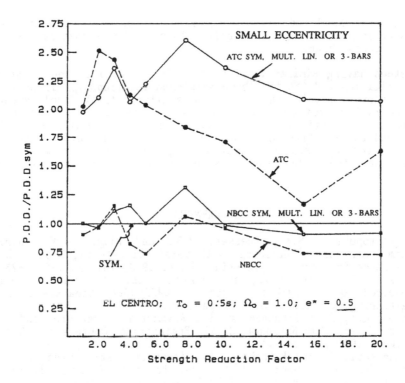

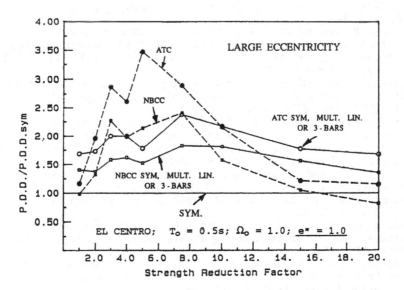

Figure 13. Comparison of asymmetric and symmetric PDD using the actual symmetric force-displacement relation: NBCC and ATC codes [38].

system for the NBCC and ATC-3 designs. It can be seen that the NBCC multilinear symmetric response is quite close to the asymmetric response in of the two cases shown.

Systems having similar total translational and rotational strengths, PC eccentricities and the two above-mentioned relationships in addition to similar linear properties should respond similarly (peak displacements and PDD). It is believed that on the basis of these force-displacement and torque-rotation relationships, the differences among models could become clearer, and the relative merits of the various codes could be quantified. Finally, better understanding would most probably lead to more adequate statically oriented design procedures for nonlinear asymmetric structures.

REFERENCES

1. Anagnostopoulos, S.A., Roesset, J.M. and Biggs, J.M., Nonlinear dynamic analysis of buildings with torsional effects, Proc. 5th World Conference Earthquake Engineering, Rome, 1973, Vol. 2, 1822-1825.
2. Batts, M.E., Berg, G.V., and Hanson, R.D.. Torsion in buildings subjected to earthquakes. Report No. UMEE 78R4, Department of Civil Engineering, University of Michigan, 1978.
3. Benbenishti, A. and Rutenberg, A., Asymmetric structures and seismic codes. Proc. 7th National Civil Engineering Conference, Tel-Aviv, December 1990 (in Hebrew).
4. Benbenishti, A. and Rutenberg, A., Asymmetric structures and seismic codes: Ductility demand. Publ. 305, Faculty of Civil Engineering, Technion - Israel Institute of Technology, Haifa, January 1991.
5. Bozorgnia, Y. and Tso, W.K., Inelastic earthquake response of buildings. Journal Structural Engineering, ASCE, 1986, 112, 383-400.
6. Bruneau, M. and Mahin, S.A., Inelastic seismic response of structures with mass or stiffness eccentricities in plan. Report UBC/EERC 87/2, Earthquake Engineering Research Center, University of California, berkeley, September 1987.
7. Bruneau, M. and Mahin, S.A., Normalizing inelastic seismic response of structures having eccentricities in plan. J. Structural Engineering, ASCE, 1990, 116, 3358-3379.
8. Bruneau, M. and Mahin, S.A., Inelastic torsional response of initially symmetric systems with lateral load resisting elements having dissimilar yield strengths. Proc. 9th European Conference Earthquake Engineering, Moscow, 1990, Vol. 6, 13-32.
9. Bruneau, M. and Mahin, S.A., A method to achieve parametric independence on the seismic inelastic response of torsionally coupled systems. Proc. 9th European Conference Earthquake Engineering, Moscow, 1990, Vol. 6, 33-42.
10. Cardona Nunez, R., Sobre la respuesta torsional de edificios de cortante. Master of Engineering Thesis, Faculty of Engineering, UNAM, Mexico, DF, 1977.
11. Chandler, A.M. and Duan, X.N., Inelastic torsional behaviour of asymmetric buildings under severe earthquake shaking. Structural Engineering Review, 1990, 2, 141-159.
12. Chandler, A.M. and Duan, X.N., Evaluation of factors influencing the inelastic seismic performance of torsionally asymmetric buildings. Earthquake Engineering and Structural Dynamics, 1991, 20, 87-95.

13. Diaz-Molina, I., Dynamic torsional behavior of inelastic systems. M. Sc. Report, Department of Civil Engineering, Carnegie Mellon University, Pittsburgh, October 1988.

14. Duan, X. and Chandler, M., Torsional coupling effects in the inelastic seismic response of structures in Europe, Proc. 9th European Conference Earthquake Engineering, Moscow, 1990, Vol. 6, 162-171.

15. Eisenberger, M. and Rutenberg, A., Seismic base isolation of asymmetric shear buildings. Engineering Structures, 1986, 8, 2-8.

16. Erdik, M.O., Torsional effects in dynamically excited structures. Ph.D. Thesis, Rice University, Houston, May 1975.

17. Esteva, L., Earthquake engineering research and practice in Mexico after the 1985 earthquake. Bull. New Zealand National Society for Earthquake Engineering, 1987, 20, 159-200.

18. Garcia-Ranz, F. and Gomez, R. (trans.), The Mexico Earthquake of Sept. 19, 1985 - Seismic Design Regulations for the 1976 Mexico Building Code. Earthquake Spectra, 1988, 4, 427-439.

19. Glück, J., Reinhorn, A. and Rutenberg, A., Dynamic torsional coupling in tall building structures. Proc. Institution of Civil Engineers 1979, 67, Pt. 2, 411-424.

20. Goel, R.K. and Chopra, A.K., Inelastic earthquake response of one-storey asymmetric-plan systems. Report UBC/EERC-90/14, Earthquake Engineering Research Center, University of California, Berkeley, October 1990.

21. Goel, R.K. and Chopra, A.K., Inelastic seismic response of one-storey asymmetric-plan systems: Effects of stiffness and strength distribution. Earthquake Engineering and Structural Dynamics, 1990, 19, 949-970.

22. Gomez, R., Ayala, G. and Jaramillo, J.D., Respuesta sismica de edificios asimetricos. Internal Report, Instituto de Ingenieria UNAM, Mexico DF, May 1987.

23. Gomez, R., Ayala, G. and Jaramillo, J.D., Respuesta sismica de edificios asimetricos, Part 2. Internal Report, Instituto de Ingenieria, UNAM, Mexico DF, June 1988.

24. Gomez, R. and Garcia-Ranz, F. (trans.), The Mexico earthquake of September 19, 1985 - Complementary technical normes: Earthquake resistant design 1987 edn. Earthquake Spectra, 1988, 4, 441-459.

25. Housner, G.W. and Outinen, H., The effect of torsional oscillations on earthquake stresses. Bull. Seismological Society of America, 1958, 48, 221-229.

26. Irvine, H.M. and Kountouris, G.E., Inelastic seismic response of a torsionally unbalanced single-story building model. Publ. R79-31, Dept. of Civil Engineering, Massachusetts Institute of Technology, Cambridge, Mass., July 1979.

27. Irvine, H.M. and Kountouris, G.E., Peak ductility demands in simple torsionally unbalanced building models subjected to earthquake excitation. Proc. 7th World Conference Earthquake Engineering, Istanbul, Vol. 4, 117-120, 1980.

28. Kan, C.L. and Chopra, A.K., Linear and nonlinear earthquake response of simple torsionally coupled systems. Report UCB/EERC-79/03, Earthquake Engineering Research Center, University of California, Berkeley, February 1979.

29. Kan, C.L. and Chopra, A.K., Torsional coupling and earthquake response of simple elastic and inelastic systems. Journal of the Structural Division, ASCE, 1981, 107, 1569-88.

30. Palazzo, B. and Fraternali, F., Seismic ductility demand in buildings irregular in plan: A new single storey nonlinear model. Proc. 9th

World Conference Earthquake Engineering, Tokyo & Kyoto, Vol. V, 43-48, 1988.

31. Rosenblueth, E., Aseismic provisions for the federal district, Mexico. Proc. 2nd. World Conference Earthquake Engineering, Tokyo, 2009-2026, 1960.

32. Rosenblueth, E., Considerations on torsion, overturning and drift limitations. Proc. SEAOC Conference, Coronado, 36-38, November 1957.

33. Rosenblueth, E. and Meli, R.L., The 1985 earthquake: Causes and effects in Mexico City. Concrete International, May 1986, 23-35.

34. Rutenberg, A., A consideration of the torsional response of building frames. Bull. New Zealand National Society for Earthquake Engineering, 1979, 12, 11-21.

35. Rutenberg, A., Eisenberger, M. and Shohet, G., Reducing seismic ductility demand in asymmetric shear buildings. Proc. 9th. European Conference Earthquake Engineering, Lisbon, 1986, 6.7/57-6.7/64.

36. Rutenberg, A. and Pekau, O.A., Earthquake response of asymmetric buildings: a parametric study. Proc. 4th Canadian Conference Earthquake Engineering, Vancouver, 1983, 271-281.

37. Rutenberg, A. and Pekau, O.A., Seismic code provisions for asymmetric structures: a re-evaluation. Engineering Structures, 1987, 9, 255-264.

38. Rutenberg, A. and Pekau, O.A., Seismic response of asymmetric structures and building codes. Unpublished seminar notes, St. Louis, September 1989.

39. Rutenberg, A., Shohet, G. and Eisenberger, M., Inelastic seismic response of code designed asymmetric structures. Publ. 303, Faculty of Civil Engineering, Technion - Israel Institute of Technology, Haifa, December 1989 (revised version of summer 1987 manuscript).

40. Sadek, A.W., and Tso, W.K., Strength eccentricity concept for inelastic analysis of asymmetrical structures. Engineering Structures, 1989, 11, 189-194.

41. Shibata, A., Onose, J. and Shiga, T., Torsional response of buildings to strong earthquake motions. Proc. 4th World Conference Earthquake Engineering, Santiago, 1969, 123-138.

42. Shiga, T. et al., Torsional response of structures to earthquake motion. Proc. U.S. Japan Seminar on Earthquake Engineering, Sendai, 1970, 156-171.

43. Shohet, G., Ductility demand in asymmetric structures. M. Sc. Thesis, Faculty of Civil Engineering, Technion, Israel Institute of Technology, Haifa, June 1986.

44. Syamal, P. and Pekau, O.A., Dynamic response of bilinear asymmetric structures. Earthquake Engineering and Structural Dynamics, 1985, 13, 527-541.

45. Tanabashi, R., Kobori, T. and Kaneta, K., Nonlinear torsional vibration of structures due to an earthquake. Bull. 56, Disaster Prevention Research Institute, Kyoto University, Kyoto, March 1962.

46. Tso, W.K. and Bozorgnia, Y., Effective eccentricity for inelastic response of buildings. Earthquake Engineering and Structural Dynamics, 1986, 14, 413-427.

47. Tso, W.K. and Sadek, A.W., Inelastic response of eccentric buildings subjected to bidirectional ground motions. Proc. 8th World Conference Earthquake Engineering, San Francisco, vol. 4, 1984, 203-210.

48. Tso, W.K. and Sadek, A.W., Inelastic seismic response of simple eccentric structures. Earthquake Engineering and Structural Dynamics, 1985, 19, 243-258.

49. Tso, W.K. and Ying, M., Additional seismic inelastic deformation caused by structural asymmetry. Earthquake Engineering and Structural

<u>Dynamics</u>, 1990, **19**, 243-258.
50. Yiaslas, I.I., Marathias, P.P. and Ikonomou, A.S., Ductility demand of one story eccentric structures subjected to seismic motion. In <u>Application of Model Analysis to External Loads</u>, ASME Publ. 1988, 150, 9-15.
51. Yamazaki, Y., Inelastic torsional response of structures subjected to earthquake ground motions. Report UBC/EERC-80/07, Earthquake Engineering Research Center, University of California, Berkeley, April 1980. Also: BRI Resaerch Paper 92, Building Research Institute, Ministry of Construction, Tsukuba, March 1981.

CODES

C1. "Tentative Provisions for the Development of Seismic Regulations for Buildings". ATC 3-06, Applied Technology Council, National Bureau of Standards, Washington, D.C., 1978.
C2. "National Building Code of Canada and Supplement". Associate Committee on the National Building Code, National Research Council of Canada, Ottawa, 1985.
C3. "Seismic Design of Concrete Structures". Comité en Euro-International du Beton, CEB, Technical Press, London, 1987.
C4. "Eurocode 8: Common Unified rules for Structures in Seismic Regions". Commission of the European Communities, Brussels, Draft, 1987.
C5. "Uniform Building code". International Conference of Building Officials, Whittier, 1979.
C6. "NEHRP Recommended Provisions for the Development of Seismic Regulations for New Buildings, Part 1 Provision". Building Seismic Safety Council, Federal Emergency Management Agency (FEMA), Washington, D.C., 1988.
C7. "New Zealand Standard: Code of Practice for General Structural Design and Design Loadings for Buildings". Standards Association of New Zealand NZS4203. Wellington, 1984.
C8. "Recommended Lateral Force Requirements and Tentative Commentary". Seismology Committee, Structural Engineers Association of California, 1975.

LIST OF CONTRIBUTORS